my **revisi⏻n** notes

CCEA GCSE

PHYSICS

Roy White

HODDER
EDUCATION
AN HACHETTE UK COMPANY

Photo on page 135 is reproduced by permission of NASA/JPL-Caltech/ Harvard-Smithsonian CfA.

Every effort has been made to trace all copyright holders, but if any have been inadvertently overlooked, the Publishers will be pleased to make the necessary arrangements at the first opportunity.

Although every effort has been made to ensure that website addresses are correct at time of going to press, Hodder Education cannot be held responsible for the content of any website mentioned in this book. It is sometimes possible to find a relocated web page by typing in the address of the home page for a website in the URL window of your browser.

Hachette UK's policy is to use papers that are natural, renewable and recyclable products and made from wood grown in sustainable forests. The logging and manufacturing processes are expected to conform to the environmental regulations of the country of origin.

Orders: please contact Bookpoint Ltd, 130 Milton Park, Abingdon, Oxon OX14 4SE. Telephone: +44 (0)1235 827720. Fax: +44 (0)1235 400454. Email: education@bookpoint.co.uk Lines are open from 9 a.m. to 5 p.m., Monday to Saturday, with a 24-hour message answering service. You can also order through our website: www.hoddereducation.co.uk

ISBN: 978-1-5104-0449-6

First published in 2017 by

Hodder Education,
An Hachette UK Company
Carmelite House
50 Victoria Embankment
London EC4Y 0DZ

www.hoddereducation.co.uk

Impression number   10 9 8 7 6 5 4
Year   2021 2020

Cover photo © Panther Media GmbH/Alamy Stock Photo

Typeset in Bembo Std Regular 11/13 by Integra Software Services Pvt. Ltd, Pondicherry, India

Printed in Spain

A catalogue record for this title is available from the British Library.

Get the most from this book

Everyone has to decide his or her own revision strategy, but it is essential to review your work, learn it and test your understanding. These Revision Notes will help you to do that in a planned way, topic by topic. Use this book as the cornerstone of your revision and don't hesitate to write in it — personalise your notes and check your progress by ticking off each section as you revise.

Tick to track your progress

Use the revision planner on pages iv–vii to plan your revision, topic by topic. Tick each box when you have:

- revised and understood a topic
- tested yourself
- practised the exam questions and gone online to check your answers

You can also keep track of your revision by ticking off each topic heading in the book. You may find it helpful to add your own notes as you work through each topic.

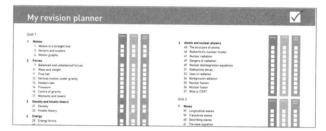

Features to help you succeed

Exam tips

Expert tips are given throughout the book to help you polish your exam technique in order to maximise your chances in the exam.

Typical mistakes

The author identifies the typical mistakes candidates make and explains how you can avoid them.

Now test yourself

These short, knowledge-based questions provide the first step in testing your learning. Go online to check your answers.

Definitions and key words

Clear, concise definitions of essential key terms are provided where they first appear.

Examples

Worked examples are given throughout the book.

Exam practice

Practice exam questions are provided for each topic. Use them to consolidate your revision and practise your exam skills. Go online to check your answers.

Online

Go online to check your answers to the questions at **www.hoddereducation.co.uk/ myrevisionnotesdownloads**.

Level coding

If you are taking GCSE Double Award Foundation-tier you need to study *only* the material with no bars.

If you are taking GCSE Double Award Higher-tier you need to study the material with no bars, plus the material with the purple H bar and the material with the orange bar.

If you are taking GCSE Chemistry Foundation-tier you need to study the material with no bars, plus the material with the green F bar , and the material with the orange bar.

If you are taking GCSE Chemistry Higher-tier you need to study all material in the book, including the material marked with the green H bar.

CCEA GCSE Physics

My revision planner

	1 REVISED	2 TESTED	3 EXAM READY

Unit 2

	REVISED	TESTED	EXAM READY

REVISED   TESTED   EXAM READY

Now test yourself answers and exam practice answers at
www.hoddereducation.co.uk/myrevisionnotesdownloads

Countdown to my exams

6–8 weeks to go

- Start by looking at the specification — make sure you know exactly what material you need to revise and the style of the examination. Use the revision planner on pages iv–vii to familiarise yourself with the topics.
- Organise your notes, making sure you have covered everything on the specification. The revision planner will help you to group your notes into topics.
- Work out a realistic revision plan that will allow you time for relaxation. Set aside days and times for all the subjects that you need to study, and stick to your timetable.
- Set yourself sensible targets. Break your revision down into focused sessions of around 40 minutes, divided by breaks. These Revision Notes organise the basic facts into short, memorable sections to make revising easier.

REVISED ☐

2–6 weeks to go

- Read through the relevant sections of this book and refer to the exam tips, typical mistakes and key terms. Tick off the topics as you feel confident about them. Highlight those topics you find difficult and look at them again in detail.
- Test your understanding of each topic by working through the 'Now test yourself' questions in the book. Look up the answers at **www.hoddereducation.co.uk/ myrevisionnotesdownloads**
- Make a note of any problem areas as you revise, and ask your teacher to go over these in class.
- Look at past papers. They are one of the best ways to revise and practise your exam skills. Write or prepare planned answers to the exam practice questions provided in this book. Check your answers online at **www.hoddereducation. co.uk/myrevisionnotesdownloads**
- Try different revision methods. For example, you can make notes using mind maps, spider diagrams or flash cards.
- Track your progress using the revision planner and give yourself a reward when you have achieved your target.

REVISED ☐

One week to go

- Try to fit in at least one more timed practice of an entire past paper and seek feedback from your teacher, comparing your work closely with the mark scheme.
- Check the revision planner to make sure you haven't missed out any topics. Brush up on any areas of difficulty by talking them over with a friend or getting help from your teacher.
- Attend any revision classes put on by your teacher. Remember, he or she is an expert at preparing people for examinations.

REVISED ☐

The day before the examination

- Flick through these Revision Notes for useful reminders, for example the exam tips, typical mistakes and key terms.
- Check the time and place of your examination.
- Make sure you have everything you need — extra pens and pencils, ruler, calculator, protractor, tissues, a watch, bottled water.
- Allow some time to relax and have an early night to ensure you are fresh and alert for the examinations.

REVISED ☐

My exams

GCSE Physics Paper 1
Date:..
Time:..
Location:...

GCSE Physics Paper 2
Date:..
Time:..
Location:...

GCSE Physics Paper 3A
Date:..
Time:..
Location:...

GCSE Physics Paper 3B
Date:..
Time:..
Location:...

1 Motion

Motion in a straight line

REVISED

The **distance** between two points is how far they are apart.

Displacement measures their distance apart *and* specifies the direction.

In the same way, **speed** is the rate at which distance travelled changes with time, but **velocity** is the rate of change of displacement with time.

$$\text{average speed} = \frac{\text{total distance travelled}}{\text{total time taken}} \qquad \text{average velocity} = \frac{\text{total displacement}}{\text{total time taken}}$$

> **displacement** is the distance between two points in a specified direction
>
> **speed** is the rate at which distance changes with time
>
> **velocity** is the rate of change of displacement with time

Suppose Jo walks around the three sides of the sports pitch shown in Figure 1.1 in a time of 250 seconds.

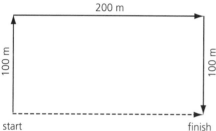

Figure 1.1

> **Exam tip**
>
> These formulae must be memorised.

She has travelled a **distance** of **400 metres**. But her final **displacement** is **200 m to the right** of the starting position.

Her **average speed** is given by:

$$\text{average speed} = \frac{\text{total distance travelled}}{\text{total time taken}} = \frac{400}{250}$$

$$= 1.6 \, \text{m/s}$$

> **average speed** is the total distance travelled divided by the total time taken

The rate of change of speed with time is defined by the equation:

$$\text{rate of change of speed} = \frac{\text{final speed} - \text{initial speed}}{\text{time taken}}$$

Rate of change of speed is measured in m/s^2.

Rate of change of speed has no direction since speed has no direction.

If the rate of change of speed is constant, then average speed can be calculated from the formula:

$$\text{average speed} = \frac{\text{initial speed} + \text{final speed}}{2}$$

> **Exam tip**
>
> These formulae must be memorised.

(H) Her **average velocity** is given by:

$$\text{average velocity} = \frac{\text{total displacement}}{\text{total time taken}} = \frac{200}{500}$$

$$= 0.8 \, \text{m/s to the right}$$

> **average velocity** is the total displacement divided by the total time taken

If the starting position and finishing position are the same, then the total displacement will be zero and hence the average velocity must be zero also.

H Acceleration is defined as the rate of change of velocity with time. The definition can be written as an equation:

$$a = \frac{v - u}{t} = \frac{\text{change in velocity}}{\text{time taken}} = \frac{\Delta v}{t}$$

or

$$v = u + at$$

where a is the acceleration in m/s^2

u is the initial (starting) velocity in m/s

v is the final velocity in m/s

t is the time taken in seconds

Δv = change in velocity in m/s

If the acceleration is constant, the average velocity can be calculated using the formula:

$$\text{average speed} = \frac{\text{initial speed} + \text{final speed}}{2}$$

A positive acceleration means the velocity is increasing, while a negative acceleration means a decreasing velocity or retardation.

Table 1.1 represents an acceleration of $2\,\text{m/s}^2$. Every second the velocity increases by 2 m/s.

Table 1.2 represents an acceleration of $-2\,\text{m/s}^2$. Every second the velocity decreases by 2 m/s.

> **acceleration** is the rate of change of velocity with time

> **Exam tip**
>
> These formulae must be memorised.

Table 1.1

Velocity/m/s	12	14	16	18	20
Time/s	0	1	2	3	4

Table 1.2

Velocity/m/s	19	17	15	13	11
Time/s	0	1	2	3	4

One of the prescribed practicals (P1) that you are required to carry out involves using simple apparatus, including trolleys, ball bearings, metre rulers, stop clocks and ramps, to investigate experimentally how the average speed of an object moving down a runway depends on the slope of the runway measured as the height of one end of the runway.

A possible approach is given below.

Prescribed practical P1

Average speed

Method

1 See Figure 1.2. Set up a ramp against a small pile of wooden blocks (or books).
2 With a ruler draw two pencil lines on the ramp, one at the top and the other at the bottom, 1.0 m apart. This is the distance, *x*.
3 Measure the height of the ramp, *h*. This is the independent variable. The dependent variable is the average speed of a trolley moving down the ramp.
4 Allow a trolley to roll down the ramp, starting from rest at the upper pencil line and finishing as its wheels reach the lower pencil line.
5 For each height, *h*, ranging from about 1 cm to about 5 cm, time this motion a total of *three* times using a stop clock and record the results in a table such as shown on page 3.
6 Calculate the average time, *t*.

→

Answers at www.hoddereducation.co.uk/myrevisionnotesdownloads

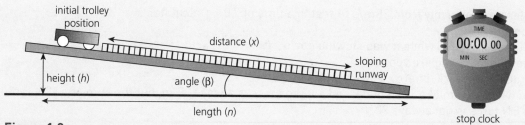

initial trolley position

distance (x)

sloping runway

height (h)

angle (β)

length (n)

TIME
00:00 00
MIN   SEC

stop clock

Figure 1.2

Results

The average speed is approximately equal to x/t.

Height (h)/cm	1	2	3	4	5
Time (t_1)/s	8.2	5.9	4.6	4.1	3.5
Time (t_2)/s	8.2	5.8	4.7	4.1	3.7
Time (t_3)/s	8.1	5.8	4.8	4.2	3.9
Average time (t)/s	8.2	5.8	4.7	4.1	3.7
Average speed (v)/cm/s	12.2	17.2	21.3	24.4	27.0

The data in the table are typical of what might be obtained.

7 Plot the graph of average speed (y-axis) against height, h (x-axis). The line of best fit is a curve through the origin of decreasing gradient. The graph shows that the average speed is **not** proportional to h, but it increases with h in some (unknown) non-linear way.

Many possible approaches to this experiment are possible. For example, the timing may be done with light gates and a data-logger, and the data analysis and graph plotting might be done using suitable computer software.

⊕ Vectors and scalars

REVISED

A **scalar** is a physical quantity that has magnitude, but not direction. Examples are distance, speed, rate of change of speed and time.

A **vector** is a physical quantity that has both magnitude and direction. Examples are displacement, velocity, acceleration and force.

For every physical quantity you encounter in your GCSE course, you need to be able to state its unit and whether it is a scalar or a vector.

> **scalar** is a physical quantity that has magnitude but not direction
>
> **vector** is a physical quantity that has both magnitude and direction

Now test yourself

TESTED

1 An athlete jogs five times around a rectangular track measuring 105 m by 150 m in a time of 850 s. Calculate:
  (a) the total distance travelled   2,550m ✓
  (b) the athlete's average speed.   3 m/s ✓
2 A sports car can increase its speed from 3 m/s to 27 m/s in 8 s. Calculate:
  (a) its rate of increase in speed, assuming it is constant   3 m/s² ✓
  (b) its average speed   15 m/s
  (c) the distance travelled during the 8 s its speed was increasing.   ~~I them~~
3 A marble takes 3.7 s to roll down a runway of length 100 cm starting from rest.
  (a) Calculate:
    (i) the marble's average speed   27.0m²
    (ii) the marble's maximum speed   54 cm/s
    (iii) the rate at which the speed of the marble was increasing.
  (b) What assumption did you make in parts (a)(ii) and (iii)?

4 A car decelerates uniformly from 28 m/s to rest in a time of 7.0 s. Calculate:
   (a) its retardation   ~4 m/s
   (b) the distance travelled while it was slowing down.   98 m
5 (a) Which of the following are vectors?
     speed, distance, work, energy, power, time
   (b) Explain why the sum of a 12 kg mass and an 8 kg mass is always 20 kg, but the sum of a 12 N force and an 8 N force is not always 20 N.

Answers online

Motion graphs

You need to be able to interpret distance–time and velocity–time graphs to solve questions of a mathematical nature. Here are some essential ideas which you must remember:

- The **gradient** of a **distance–time graph** represents an object's **speed**.
- The **gradient** of a **speed–time graph** represents an object's **rate of change of speed**.
- The **area** between a **speed–time graph** and the time axis represents the **distance moved**.
- The **gradient** of a **displacement–time graph** represents an object's **velocity**.
- The **gradient** of a **velocity–time graph** represents an object's **acceleration**.
- The **area** between a **velocity–time graph** and the time axis represents the **displacement**.

Examples

1 A speed–time graph for a tractor is shown in Figure 1.3. Calculate:
   (a) the distance travelled between times $t = 0$ s and $t = 30$ s
   (b) the rate of change of speed between times $t = 15$ s and $t = 20$ s.

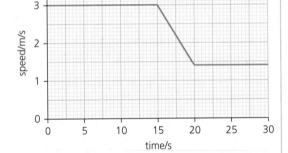

Figure 1.3

Answer

(a) distance = area between graph and the time axis.
   Divide graph into two rectangles and a triangle as shown in Figure 1.4.

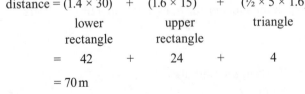

$$\text{distance} = \underset{\substack{\text{lower}\\\text{rectangle}}}{(1.4 \times 30)} + \underset{\substack{\text{upper}\\\text{rectangle}}}{(1.6 \times 15)} + \underset{\text{triangle}}{(½ \times 5 \times 1.6)}$$

$$= \quad 42 \quad + \quad 24 \quad + \quad 4$$

$$= 70 \, \text{m}$$

(b) rate of change of speed = gradient of speed–time graph

$$= \frac{1.4 - 3.0}{20 - 15} = \frac{-1.6}{5} = -0.32 \, \text{m/s}^2$$

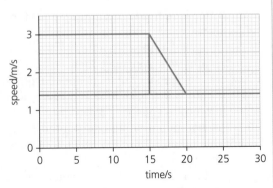

Figure 1.4

2 A pupil runs a race and the velocity–time graph for the race is shown in Figure 1.5 .Calculate:
   (a) the length of the race
   (b) the acceleration during the first 4 seconds.

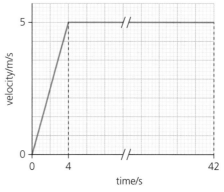

Figure 1.5

Answer

(a) displacement = area between v–t graph and the time axis

displacement = area of triangle + area of rectangle
$$= \frac{1}{2}(4 \times 5) \qquad + \qquad (38 \times 5)$$
$$= 10 \qquad\qquad + \qquad 190$$
$$= 200\,\text{m}$$

(b) acceleration = gradient of velocity–time graph

$$\text{acceleration} = \frac{\text{change in velocity}}{\text{change in time}} = \frac{5-0}{4-0} = 1.25\,\text{m/s}^2$$

3 Jim runs home from school each day. The graph in Figure 1.6 shows part of his journey.
   (a) How far from school is Jim after 15 seconds?
   (b) What is Jim's steady speed during the first 15 seconds of his motion?
   (c) Describe the motion during the last 10 seconds of the journey.
   (d) Calculate Jim's average speed for the entire 25 seconds of the journey.

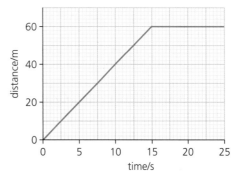

Figure 1.6

Answer

(a) At $t = 15\,\text{s}$, distance to school is 60 m.

(b) $\text{speed} = \dfrac{\text{distance}}{\text{time}} = \dfrac{60}{15} = 4\,\text{m/s}$

(c) Jim is stopped for the last 10 s.

(d) $\text{average speed} = \dfrac{\text{total distance}}{\text{total time}} = \dfrac{60}{25} = 2.4\,\text{m/s}$

You should also remember that the average velocity along the straight part of a velocity–time graph is the average of the velocities at the start and end of the straight line. This is tested in question 4 in the exam practice at the end of this chapter.

Exam practice

1 A soldier marches 15 metres due east, then turns and marches 30 metres due west. He turns again and marches 30 metres due east. The total time taken is 25 seconds. Calculate:
   (a) his final displacement from his starting position   15m ✓
   (b) the total distance marched   75 ✓
   (c) his average speed over the 25 seconds   3m/s ✓
   (d) his average velocity.   0·6m/s due east ✓

→

2 A graph of velocity against time for a golf ball is shown in Figure 1.7.

(a) Use the graph to find the deceleration of the ball $0.4\ m/s$

(b) How far does the ball travel in 6 seconds? $7.2\ m$

$g = \dfrac{y}{x}$

$y = \dfrac{2.4}{6}$

$y = 0.4\ m/s$ ✓

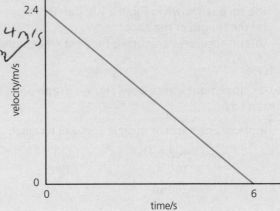

Figure 1.7

3 A father runs a 200 metre race against his son. Both start from the same position. The father gives his son a start by not beginning to run until the boy is some distance ahead of him. A distance–time graph is shown for both father and son in Figure 1.8.

(a) How far ahead was the son when the father began to run? $60m$ ✓

(b) How far from the start did the father overtake his son? $120m$ ✓

(c) When the father finished the race, how far behind was the son? $40m$

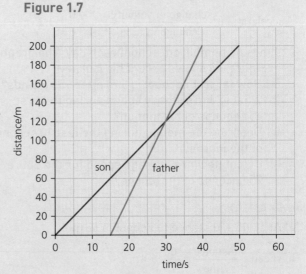

Figure 1.8

4 Car A and car B are in a race and their velocity–time graphs are shown in Figure 1.9.

(a) How far has car A travelled after 12 seconds? $36m$ ✓

(b) How far apart are the two cars after 12 seconds? $24\ m$ ✓

(c) Calculate the average velocity of car B.

$avg\ velocity = \dfrac{60}{12}$

$= 5\ m/s$ ✓

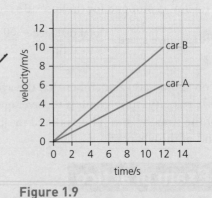

Figure 1.9

5 Maureen cycles to school each day. The graph in Figure 1.10 illustrates her journey. Use the graph to calculate her speed for the last part, BC, of the journey.

$g = \dfrac{y}{x}$

$g = \dfrac{60}{30}$

$g = 2\ m/s$ ✓

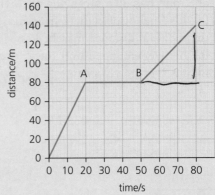

Answers online

Figure 1.10

ONLINE

2 Forces

Balanced and unbalanced forces

Force is a vector — it has both size and direction. The size of the force is measured in **newtons** (N). In diagrams a force is represented by an arrow.

Equal forces acting in opposite directions are balanced. **Balanced forces** do not change the velocity of an object. This is summed up in **Newton's first law**:

A body stays at rest, or if moving it continues to move with uniform velocity, unless an unbalanced force makes it behave differently.

> **balanced forces** occur when there is no resultant force on an object

> **Exam tip**
>
> This law must be memorised.

Practical

Newton's first law

To demonstrate Newton's first law we use a linear air track and blower (to minimise friction), a glider and interrupt card, two light gates, a data logger and a computer (see Figure 2.1).

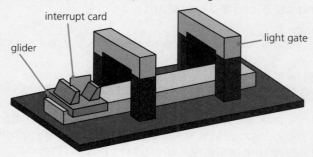

Figure 2.1

Method

1 Set the linear air track on a flat bench and adjust the feet on the air track to make sure that it is level.
2 Measure the length of the interrupt card and enter this in the data logger.
3 Connect up the light gates so that they measure the velocity of the glider at two points.
4 Give the glider a gentle push so that it passes through both light gates.
5 Confirm by looking at the results that the velocity does not change between the two positions of the light gate and so the glider is obeying Newton's first law of motion.
6 Repeat for other velocities and positions of the light gates.

Unbalanced forces will change the velocity of an object. Since velocity involves both speed and direction, unbalanced forces can make an object speed up, slow down or change direction. This is summed up in **Newton's second law**:

> The acceleration of a body is directly proportional to the force applied to it and inversely proportional to the object's **mass**.

Newton's second law can be written as an equation:

force = mass × acceleration

$$F = m \times a$$

Friction is a force which always opposes motion.

unbalanced forces occur when there is a resultant force causing an acceleration

mass is measured in kg and shows the amount of matter in an object

Exam tip

The equation $F = ma$ and the definition of friction must be memorised.

Practicals

Relationship between force, mass and acceleration

Apparatus

- runway
- trolley
- string
- double interrupt mask
- light gate and data logger
- pulley masses
- balance

Experiment 1: Keeping the mass being accelerated constant

1 Prepare a table for results as shown on page 9.
2 See Figure 2.2. To compensate for friction, tilt the runway until the trolley moves with a constant speed after it is given a gentle push.
3 Screw the clamped pulley to the end of the bench.
4 Attach a length of string to connect the end of the trolley to a slotted mass carrier and pass the string over the clamped pulley.
5 Position the light gate in such a way that the mask on top of the trolley passes through it without hitting anything and passes through it before the masses on the end of the string hit the ground.
6 Use the light gate to measure the acceleration of the trolley for various masses on the mass carrier from 100 g to 600 g, repeating each measurement two times and taking an average. Remember, each 100 g mass is equivalent to 1 N.
7 Find the weight of the masses and record this as the value for the resultant force F.
8 Plot a graph of acceleration (y-axis) versus force (x-axis).

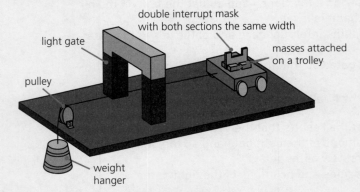

Figure 2.2 Diagram of apparatus used to show that $F = ma$

Resultant force/N	1.0	2.0	3.0	4.0	5.0	6.0
Acceleration/m/s²	0.80	1.59	2.40	3.21	4.10	4.80
Acceleration/m/s²	0.80	1.61	2.40	3.19	3.90	4.80
Mean acceleration/m/s²	0.80	1.60	2.40	3.20	4.00	4.80

A graph of mean acceleration (vertical axis) against resultant force (horizontal axis) gives a straight line through the origin (Figure 2.3), showing that the acceleration is directly proportional to the resultant force, when the mass being accelerated is constant.

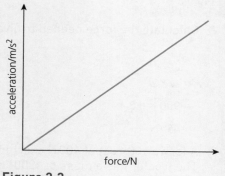

Figure 2.3

Experiment 2: Keeping the accelerating force constant

1 Prepare a table for results as shown below.
2 See Figure 2.2. To compensate for friction, tilt the runway until the trolley moves with a constant speed after it is given a gentle push.
3 Screw the clamped pulley to the end of the bench.
4 Attach a length of string to connect the end of the trolley to a slotted mass carrier and pass the string over the clamped pulley.
5 Position the light gate in such a way that the mask on top of the trolley passes through it without hitting the light gate and passes through it before the masses on the end of the string hit the ground.
6 Choose a suitable value for the driving force provided by the falling weights, e.g. 500 g (5 N).
7 Use the light gate to measure the acceleration of the trolley for various masses of trolley by either adding slotted masses to the trolley or stacking trolleys on top of each other.
8 Repeat each measurement and calculate the mean acceleration.
9 Plot graphs of (a) acceleration (y-axis) versus mass of trolley (x-axis) and (b) acceleration against 1/mass.

Mass/kg	1.0	2.0	3.0	4.0	5.0	6.0
Acceleration/m/s²	2.40	1.20	0.79	0.62	0.48	0.39
Acceleration/m/s²	2.40	1.20	0.81	0.58	0.48	0.41
Mean acceleration/m/s²	2.40	1.20	0.80	0.60	0.48	0.40

A graph of mean acceleration (vertical axis) against 1/mass (horizontal axis) yields a straight line through the origin (Figure 2.4) showing that the acceleration is **inversely proportional** to mass (or directly proportional to 1/mass) when the accelerating force is constant.

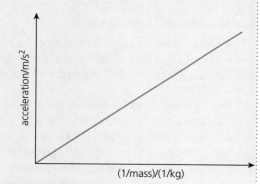

Figure 2.4

1 Calculate the force needed to give a car of mass 900 kg an acceleration of 0.5 m/s².

Answer

$F = m \times a$

$\quad = 900 \times 0.5$

$\quad = 450 \, N$

2 A forward thrust of 400 N exerted by a tractor enables it to travel along a lane at constant velocity. The tractor has a mass of 1200 kg. Calculate the thrust required to accelerate the tractor at 1.5 m/s².

Answer

The phrase 'at constant velocity' is a clue to use Newton's first law. If the thrust exerted by the engine is 400 N, there must be an equal and opposite force of 400 N due to friction on the tractor. To calculate the force to accelerate the tractor we should draw a force diagram (Figure 2.5).

friction, 400 N

thrust, F

Figure 2.5

The unbalanced force is thrust minus friction, $F - 400$ N.

unbalanced force = mass × acceleration

$(F - 400) = 1200 \times 1.5$

$F - 400 \ = 1800$

$\quad\quad F = 2200 \, N$

Now test yourself

1 A car accelerates at 3.0 m/s² along a road. The mass of the car is 1200 kg and all the resistive forces add up to 400 N. Calculate the forward thrust exerted by the car's engine. ~~4000N~~

2 Calculate the force of friction on a car of mass 1100 kg if it accelerates at 2 m/s² when the engine force is 3000 N.

3 The upward drag force on a parachutist of mass 60 kg is 480 N. Calculate her acceleration.

4 A forward thrust of 400 N exerted by a speedboat engine enables the speedboat to go through the water at a constant speed (Figure 2.6). The speedboat has a mass of 500 kg. Calculate the thrust required to accelerate the speedboat at 2 m/s².

5 A car of mass 1200 kg accelerates at 3 m/s² along a road. Calculate the forward thrust exerted by the car engine if all resistive forces add up to 400 N.

6 A cyclist has a mass of 65 kg. When she provides a thrust of 100 N to her bicycle, she accelerates at 1.0 m/s². When the thrust is 140 N, the acceleration is 1.5 m/s². The friction force is constant. Calculate the mass of the bicycle and the size of the friction force.

thrust 400N

mass 500 kg

400 N

F

Figure 2.6

Answers online

Mass and weight

Mass is defined as the **amount of matter in a body**. Mass is measured in kilograms (kg). It is another example of a scalar quantity.

Weight is the **force of gravity on an object**. Since objects close to the Earth all experience the same acceleration, the **acceleration of free fall**, we can apply Newton's second law. In our case, the force is exerted by the Earth:

weight = force of gravity = mass × acceleration due to gravity

$$W = m \times g$$

Near the surface of the Earth, there is a gravitational force of **10 N** *on each* **1 kg** of mass. We say that the Earth's **gravitational field strength, g,** is 10 N/kg.

The Moon is smaller than the Earth and pulls objects towards it less strongly. On the Moon's surface the value of g is 1.6 N/kg.

In deep space, far away from the planets, there are no gravitational pulls, so g is zero, and therefore everything is weightless.

The size of g also gives the gravitational acceleration, because from Newton's second law:

$$acceleration = \frac{force}{mass}$$

or

$$g = \frac{weight}{mass}$$

So an alternative unit for g is m/s^2.

> **weight** is the force of gravity on an object
>
> **acceleration of free fall** is the acceleration of an object towards the surface of the Earth when the only force on it is the gravitational force

> **Exam tip**
>
> This equation must be memorised.

> **gravitational field strength, g** is the gravitational force on an object of mass 1 kg close to the surface of a planet

Now test yourself

7 Describe three differences between mass and weight.
8 A parachutist of weight 620 N falls vertically through the air at a constant speed of 5 m/s.
  (a) State the resultant force on the parachutist.
  (b) State the size and direction of the frictional force on the parachutist.
9 Explain why there must be a resultant force on an object moving in a circular track at a steady speed.
10 Deep in space (where friction can be taken as zero), an astronaut throws a hammer. It leaves her hand with a speed of 3 m/s. Describe its subsequent motion. On what law of physics does your description depend?

Answers online

Free fall

Galileo showed that two lead balls of different diameter hit the ground at the same instant when dropped from the top of the leaning tower of Pisa.

All bodies in the absence of air resistance fall at the same rate of 10 m/s^2 near the surface of the Earth. It is a common misconception to think that a more massive object falls faster than a less massive one. It is true that there is a greater force on the more massive object but the acceleration, which is the ratio of force to mass, will be the same for both bodies:

$$a = \frac{F}{m} \text{ or in this case } g = \frac{W}{m}$$

H This means that if there is no air resistance, as in Figure 2.7, the speed of a falling object will increase by 10 m/s every second, i.e. its acceleration is 10 m/s².

Now test yourself

11 Julie said, 'My weight is 35 kg.' What is wrong with this statement and what do you think her weight really is?

12 A ball bearing is gently dropped into a tall cylinder of oil, which resists its motion. Describe what will happen to the ball bearing.

13 An astronaut standing on the surface of the Moon releases a hammer and a feather from the same height. What will happen and why?

14 Why does a parachute slow down a falling parachutist?

Answers online

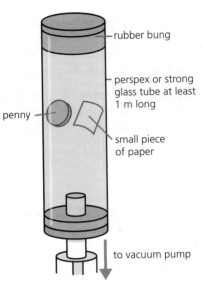

rubber bung

perspex or strong glass tube at least 1 m long

penny

small piece of paper

to vacuum pump

Figure 2.7

H Vertical motion under gravity

REVISED

When a body is thrown vertically upwards, its motion is opposed by the force of gravity. The velocity of the body will *decrease* by 10 m/s in each and every subsequent second until its vertical velocity is zero. The body has experienced negative acceleration, more often referred to as **deceleration** or **retardation**.

Example

A ball is thrown vertically upwards with an initial velocity of 50 m/s. How long will the ball take to reach the top of its motion?

Answer

Initial velocity $u = 50$ m/s

Final velocity $v = 0$ m/s

Time of vertical motion $= t$

Acceleration $= a = -10$ m/s²

$$a = \frac{v - u}{t}$$

$$-10 = \frac{0 - 50}{t}$$

$$t = \frac{-50}{-10}$$

$$t = 5 \text{ s}$$

Hooke's law states that the extension is directly proportional to the applied load, up to a limit known as the limit of proportionality

natural length is the length of a spring when no stretching or compressing forces are applied

extension is the difference between the stretched length of a spring and its natural length

Hooke's law

Hooke's law describes the behaviour of metal wires and springs when they are stretched or compressed. The difference between its extended length and its **natural length** is called its **extension**.

> **Hooke's law** states that the extension is directly proportional to the applied load, up to a limit known as the limit of proportionality
>
> **natural length** is the length of a spring when no stretching or compressing forces are applied
>
> **extension** is the difference between the stretched length of a spring and its natural length

Prescribed practical P2

Hooke's law

Apparatus

Figure 2.8 shows the apparatus.

Method

1 Measure the natural length of the spring with a metre ruler.
2 Add a 100 gram (weight = 1.0 N) mass hanger.
3 Measure the extended length of the spring.
4 Calculate and record the extension.
5 Add an additional 100 gram mass.
6 Repeat the measurements and record the results in a table, as shown below.
7 Draw a graph of load/N on the y-axis against extension/cm on the x-axis.

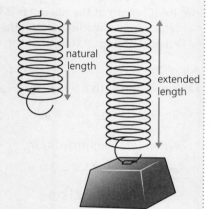

Figure 2.8

Results

Load/N	0.0	1.0	2.0	3.0	4.0	5.0
Total length/cm	4.5	6.1	7.7	9.3	10.9	12.5
Extension/cm	0.0	1.6	3.2	4.8	6.4	8.0

Figure 2.9 shows the graph produced. It should show a straight line through the origin. This is part of the region (AB) in which Hooke's law is obeyed. If you were to continue to add slotted masses and take measurements, the graph would go beyond the limit of proportionality and eventually cease to be a straight line and curve as shown.

In the straight line region the spring is **elastic**. This means it will return to its original length when the load is removed. In the curved region the spring is **plastic**. In the plastic region it will no longer return to its original length when the load is removed.

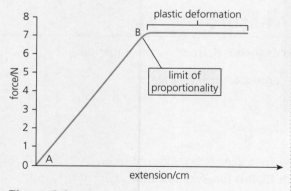

Figure 2.9

> **elastic** materials return to their original length when the stretching force is removed
>
> **plastic** materials do not return to their original length when the stretching force is removed

Hooke's law equation

Hooke's law can be written as an equation:

$$F = ke$$

where F is the applied force in N
e is the extension in cm (or mm)
k is the spring constant in N/cm or N/mm and is equal to the gradient of the force–extension graph

(Note that k is sometimes called the stiffness constant or Hooke's law constant.)

Example

A spring has a natural length of 8 cm. When loaded with a 5 N weight the total length of the spring is 23 cm.
(a) What weight would stretch the spring to a total length of 14 cm?
(b) What is the total length of the spring when the load is 7 N?

Answer

(a) Extension with a 5 N weight = 23 – 8 = 15 cm

So, $k = \dfrac{F}{e} = \dfrac{5\,N}{15\,cm} = 0.3333\,N/cm$

When total length = 14 cm, extension, e = 14 – 8 = 6 cm

$F = ke = 0.3333 \times 6 = 2\,N$

(b) $e = \dfrac{F}{k} = \dfrac{7}{0.3333} = 21\,cm$

total length = natural length + extension = 8 + 21 = 29 cm

Now test yourself

TESTED

15 State Hooke's law.
16 Explain what is meant by elastic and plastic deformation.
17 Explain what is meant by the limit of proportionality.

Answers online

Pressure

REVISED

Pressure is defined by the equation:

$$\text{pressure} = \frac{\text{force}}{\text{area}}$$

or

$$P = \frac{F}{A}$$

where force, F, is measured in N,
area, A, is measured in mm^2, cm^2 or m^2, and
pressure, P, is measured in N/mm^2, N/cm^2 or N/m^2

Note that the unit N/m^2 was renamed the **pascal (Pa)** after the French scientist of that name.

Table 2.1 lists some common situations where knowledge of physics helps you to understand what is happening.

pressure is the force per unit area on a surface

pascal is the SI unit of pressure and is the same as 1 N/m^2

Table 2.1

Situation	Comment
A wise chef sharpens a carving knife before cutting a joint of meat.	A sharp knife has a tiny area of contact with the meat, so that even a small force can produce a very large pressure. This makes cutting the joint easier.
The blade of an ice skate is sharp.	The sharp blade causes a large pressure on the ice, producing a fine layer of water between blade and ice. The water reduces friction and makes skating effortless.
A high stiletto heel can cause considerable damage to a wooden floor.	Much of the woman's weight is borne by the tiny area of the heel. This causes huge pressure on the floor. If the floor is quite soft (like wood), this causes damage.
In some places houses must be built on rafts made of concrete.	If the surrounding ground is soft there is a danger that the weight of the house will cause it to sink. The large area of the concrete spreads the weight of the house, reduces the pressure and prevents sinking.

Example

Figure 2.10 shows three stacked building blocks. Each block has a weight of 225 N.

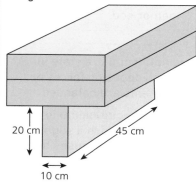

20 cm   45 cm

10 cm

Figure 2.10

(a) With the help of the measurements in Figure 2.10, calculate the pressure exerted on the ground. Give your answer in pascals.

(b) It is possible to stack the blocks on top of one another so that the pressure exerted on the ground is less than the value you have calculated. Suggest how the blocks should be arranged to give this smaller pressure.

Answer

(a) total weight $= 3 \times 225 = 675\,\text{N}$

area in contact with ground $= 0.1\,\text{m} \times 0.45\,\text{m}$

$$= 0.045\,\text{m}^2$$

$$\text{pressure} = \frac{F}{A} = \frac{675}{0.045} = 15\,000\,\text{Pa}$$

(b) Have the 20 cm × 45 cm surface in contact with the ground and the other two blocks on top of it.

TESTED ☐

Now test yourself

18 The tip of a drawing pin has an area of $1.0 \times 10^{-8}\,\text{m}^2$. Find the pressure exerted if the force applied to it is 10 N.
19 A car has a weight of 9000 N. The tyre pressure is 18 N/cm². Calculate the area of each tyre in contact with the ground.
20 A large concrete cube is of side 0.9 m. If it exerts a pressure of 61 000 Pa, calculate the cube's weight.

Answers online

Centre of gravity

REVISED ☐

The **centre of gravity** of an object is the point through which the entire weight of that object may be thought to act.

Only for regularly shaped objects is the centre of gravity, G, at the centre of the object. Centre of gravity is often wrongly thought of as the point where the object would balance.

The centre of gravity of a rectangular lamina, for example, is at its geometric centre (the point where its diagonals cross).

The idea of centre of gravity is closely linked with our notion of **stability** and **equilibrium**.

A ball on flat ground is in **neutral equilibrium**. When gently pushed the ball rolls, keeping its centre of gravity at the same height above the point of contact with the ground (Figure 2.11).

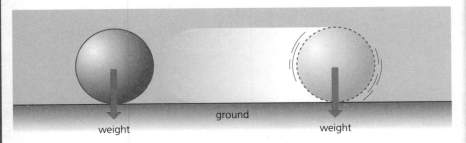

Figure 2.11

A tall radio mast is in **unstable equilibrium**. A small push from the wind will cause it to topple, bringing its centre of gravity closer to the ground. To prevent the mass toppling, it is stabilised with strong cables.

A car on the road is in **stable equilibrium**. If the car is tilted, the centre of gravity rises. Provided the weight, acting from the centre of gravity, passes through the wheel base, the car will not topple over.

Modern agricultural tractors are designed with a big wheel base to allow farmers to plough steeply sloping ground safely.

> **centre of gravity** is the point through which the entire weight of a body appears to act

> **Exam tip**
>
> This definition must be memorised.

> **equilibrium** is a state in which opposing forces or moments are balanced
>
> **neutral equilibrium** is a state in which a slight displacement causes the object neither to move very far from its original position nor return to it (e.g. a ball sitting on a bench)
>
> **unstable equilibrium** is a state in which a slight displacement causes the object to move a long way from its original position (e.g. a pencil balanced on its end
>
> **stable equilibrium** is a state in which a slight displacement causes the object to return to its original position (e.g. a traffic cone standing on its wide base)

> **Exam tip**
>
> Remember that for maximum stability:
> ● the centre of gravity should be as low as possible
> ● the area of the base should be as large as possible.

Moments and levers

Opening a door, pushing a wheelbarrow, cutting with scissors and using a nutcracker are all examples of the application of **moments** or turning forces. So what is a moment?

The moment of a force about a pivot is the product of the force and the perpendicular distance to the pivot (Figure 2.12). If the distance is in cm, then the moment is in N cm; if the distance is in metres then the moment is in N m. Moments have a direction — they are either **clockwise** or **anticlockwise**.

> the **moment** about a point is the product of a force and its perpendicular distance from the point

moment = force × distance to pivot

N cm   N   cm

The moment in Figure 2.12 is clockwise.

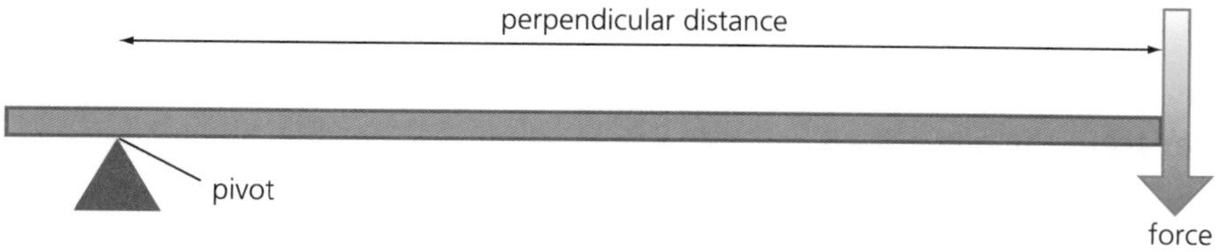

> the **principle of moments** states that when a lever is balanced, the sum of the clockwise moments about any point equals the sum of the anticlockwise moments about the same point

perpendicular distance

pivot

force

Figure 2.12

The behaviour of levers is summed up in a law known as the **principle of moments**. This law states:

> When a lever is balanced, the sum of the clockwise moments about any point equals the sum of the anticlockwise moments about the same point.

Exam tip

This law must be memorised.

Prescribed practical P3

Principle of moments

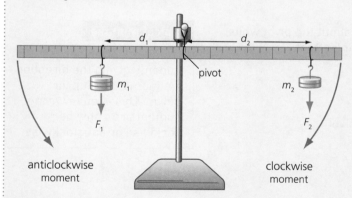

d_1 pivot d_2

m_1 m_2

F_1 F_2

anticlockwise moment   clockwise moment

Figure 2.13

Method

1 Suspend and balance a metre ruler at the 50 cm mark using twine (see Figure 2.13).
2 Adjust the position of the twine so that the rule does not rotate.
3 Hang unequal masses, m_1 and m_2 (100 g slotted masses), from either side of the metre ruler as shown in Figure 2.13.
4 Adjust the position of the masses until the metre ruler is balanced (in equilibrium) again.
5 Gravity exerts forces F_1 and F_2 on the masses m_1 and m_2. Remember that a 100 g slotted mass is equivalent to 1 N.

2 Forces

6 Record the results in a table like the one below and repeat for other loads and distances.

m_1/g	F_1/N	d_1/cm	$F_1 \times d_1$/N cm	m_2/g	F_2/g	d_2/cm	$F_2 \times d_2$/N cm

- Force F_1 is trying to turn the metre ruler anticlockwise, and its moment is $F_1 \times d_1$.
- Force F_2 is trying to turn the metre ruler clockwise — its moment is $F_2 \times d_2$.
- When the metre ruler is balanced (i.e. in equilibrium), the results should show that the anticlockwise moment $F_1 \times d_1$ equals the clockwise moment $F_2 \times d_2$.

Another important consequence of the fact that the metre ruler is in equilibrium is that the forces acting on the ruler in any direction *must* balance. The **upward forces** must balance the **downward forces**. This idea is useful when doing problems.

Exam tip

At GCSE level you will never be asked to solve problems involving more than three forces.

Example

A boy, weighing 300 N, sits 0.9 m away from the pivot of a see-saw, as shown in Figure 2.14.

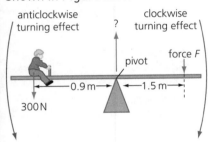

Figure 2.14

(a) What force 1.5 m from the pivot is needed to balance the see-saw?
(b) Find the size of the upward force exerted by the pivot.

Answer

(a) The force F exerts a clockwise turning effect about the pivot while the boy's weight exerts an anticlockwise turning effect.

By the principle of moments:

clockwise moment = anticlockwise moment

$F \times$ distance from pivot = $300\,\text{N} \times$ distance from pivot

$F \times 1.5\,\text{m} = 300\,\text{N} \times 0.9\,\text{m}$

$F = \dfrac{300 \times 0.9}{1.5} = 180\,\text{N}$

Typical mistake

Examiners often see students giving the direction of a moment as 'up' or 'down'. This is quite wrong. Moments can only be clockwise or anticlockwise.

To summarise:
- balanced forces — if the object is stationary it will remain stationary; if it is moving it will carry on moving at the same speed and in the same direction
- unbalanced forces — an unbalanced force on an object causes the object to accelerate, according to $F = ma$.

Exam practice

H

1 (a) State the mathematical equation linking force, mass and acceleration. [1]
   (b) An aircraft of mass 2000 kg lands at an airport. As it travels along the runway, friction forces on the aircraft add up to 8000 N. The speed of the aircraft reduces to 5 m/s in 9 s.
   (i) Calculate the aircraft's deceleration. [1]
   (ii) Calculate the speed of the aircraft when its wheels first touch the runway. [2]

2 When the resultant force on a vehicle is 3000 N, its acceleration is 4 m/s². Calculate the acceleration of the vehicle when the engine force is 3600 N and the friction force is 1725 N. [2]

3 International Cycle Union rules require racing bicycles to have a mass of 6.8 kg or more. A cyclist of mass 53 kg can produce an acceleration of 2 m/s² when the resultant force on the bicycle is 120 N. Is the mass of the bicycle within the limit set by the ICU? [2]

4 (a) Calculate the weight of an object of mass 70 kg on Earth. [2]
   (b) The same object is taken to the Moon, where g = 1.6 m/s². Calculate:
   (i) its mass and
   (ii) its weight on the Moon. [3]
   (c) On another planet, a mass of 12 kg weighs 105.6 N. Calculate the value of g on this planet. [2]
   (d) Comment on the units for g. [2]

5 A bullet, fired vertically upwards from a pistol, rises to a maximum height of 1875 m above a planet in a time of 25 seconds.
   (a) Calculate the average velocity of the bullet during this time. [3]
   (b) Using your answer to part (a), or otherwise, calculate the maximum velocity of the bullet. [3]
   The bullet then takes another 25 seconds to fall back to the planet's surface.
   (c) What is the average velocity of the bullet over the entire distance covered? [1]
   (d) Give a reason for your answer to part (c). [1]
   (e) Determine the acceleration due to gravity on the planet. [3]

6 The friction force opposing the motion of a locomotive of mass 25 000 kg is 100 000 N.
   (a) What forward force must the locomotive provide if it is to travel along a straight, horizontal track at a steady speed of 1.5 m/s? [1]
   (b) What is the acceleration of the locomotive if the forward force increases to 175 000 N and the friction force is unchanged? [3]

7 An elephant's mass is 5500 kg. The total area of its feet in contact with the ground is 0.14 m². Calculate the pressure exerted on the ground. [5]

8 The side of a large lorry measures 15 m × 2 m (Figure 2.15). On a windy day the pressure on one side is 5 kPa. Calculate the force on that side of the lorry. [5]

Figure 2.15

9 A wheelbarrow and its load together weigh 600 N. The distance between the pivot and the wheelbarrow's centre of gravity is 75 cm. The distance between the handles and the pivot is 225 cm (Figure 2.16).

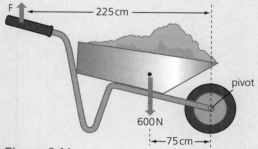

Figure 2.16

   (a) Calculate the size of the smallest force, F, needed to lift the wheelbarrow at the handles. [4]
   (b) Calculate the force acting through the pivot. [2]

10 The centre of gravity of an 80 cm snooker cue is 15 cm from its thick end (Figure 2.17).

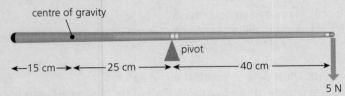

centre of gravity

pivot

←15 cm→ ←—— 25 cm ——→ ←———— 40 cm ————→

5 N

Figure 2.17

The cue balances on a pivot 40 cm from its thick end when a force of 5 N is applied to the thin end.

(a) Calculate the moment of the 5 N force about the pivot and state the direction in which it acts. [4]

(b) Calculate the weight of the snooker cue. [4]

11 A uniform plank of weight 150 N and length 4 m rests on a pivot at one end. It is balanced by an upward force, F, which acts at 3 m from the pivot. Calculate the size of force F and the size and direction of the force acting through the pivot.

Answers online

ONLINE

3 Density and kinetic theory

Density

REVISED

The **density** of a material is defined as the mass per unit volume.

It is calculated using the formula:

$$\text{density} = \frac{\text{mass}}{\text{volume}}$$

The unit of density is kilogram per cubic metre (kg/m³). Sometimes you will also see the unit gram per cubic centimetre (g/cm³).

> **density** is the mass of an object divided by its volume

Example

Taking the density of mercury as 14 g/cm³, find:
(a) the mass of 7 cm³ of mercury
(b) the volume of 42 g of mercury.

Answer

(a) $\text{density} = \dfrac{\text{mass}}{\text{volume}}$

$14 = \dfrac{\text{mass}}{7}$

$\text{mass} = 14 \times 7 = 98\,\text{g}$

(b) $\text{density} = \dfrac{\text{mass}}{\text{volume}}$

$14 = \dfrac{42}{\text{volume}}$

$\text{volume} = \dfrac{42}{14} = 3\,\text{cm}^3$

Measuring density

Ⓕ▶ Prescribed practical P4

Mass and volume

This practical requires you to investigate the relationship between the mass and volume of liquids and regular solids.

To determine the density of a substance you need to measure its mass and its volume.

1 Liquids

Apparatus

- digital balance
- measuring cylinder
- liquid

Method

1 Find the mass of a dry, empty, graduated cylinder using a digital balance (see Figure 3.1).
2 Pour the liquid under test into the cylinder and measure the volume.
3 Find the mass of the cylinder and liquid (see Figure 3.1).
4 Subtract the mass of the empty cylinder from the combined mass of cylinder and liquid.

→

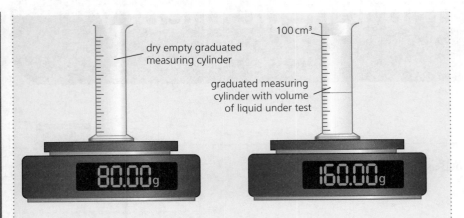

Figure 3.1 Finding the density of a liquid

5 Find the density of the liquid by dividing the mass of the liquid by the volume of the liquid.

6 For reliability, repeat the experiment and average the calculated densities.

2 Regularly shaped object

Apparatus

● digital balance
● ruler

Method

1 Find the mass of the object using a digital balance.

2 Find its dimensions with a ruler (or digital callipers) and use the appropriate formula to determine its volume.
   For example:
   volume of a rectangular block = length × breadth × height
   volume of a cylinder = π × radius² × height

3 Find the density by dividing the mass by the volume.

3 Irregularly shaped object

If the shape of the object is too irregular for the volume to be determined using formulae, then a displacement method is used, as shown in Figure 3.2.

As before, the mass is found using a digital balance and the density calculated as outlined previously.

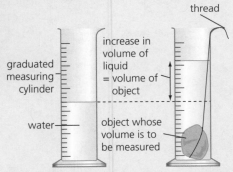

Figure 3.2 Measuring the volume of a small irregularly shaped object

Graphical treatment of density

The gradient of a mass–volume graph for different materials represents the density of a particular material (Figure 3.3).

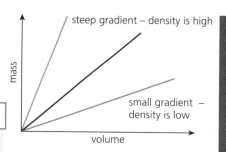

Figure 3.3

Now test yourself

TESTED

1 A substance of mass 75.6 g has a volume of 6.0 cm³. Calculate its density.
2 Aluminium has a density of 2.7 g/cm³.
   (a) What is the mass of 20 cm³ of aluminium?
   (b) What is the volume of 54 g of aluminium?
3 Calculate the mass of air in a room of dimensions 10 m by 5 m by 3 m, if air has a density of 1.26 kg/m³.
4 A stone of mass 60 g is lowered into a measuring cylinder causing the liquid level to rise from 15 cm³ to 35 cm³. Calculate the density of the stone in g/cm³.
5 The capacity of a petrol tank in a car is 0.08 m³. Calculate the mass of petrol in a full tank if the density of petrol is 800 kg/m³.
6 The mass of an evacuated steel container, of internal volume 1000 cm³, is 350 g. The mass of the steel container when full of air is 351.2 g. Calculate the density of air.

Answers online

Kinetic theory

REVISED

Table 3.1 summarises the properties of solids, liquids and gases.

Table 3.1 Explaining the variation in density of solids, liquids and gases using kinetic theory

> **kinetic theory** explains the properties of solids, liquids and gases according to the arrangement and motion of their molecules

Solids	Liquids	Gases
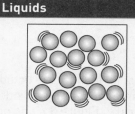	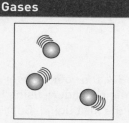	
• Molecules are packed close together.	• The molecules are **close together** but not as close as they are in solids.	• There are large distances between the molecules.
• The molecules vibrate about fixed positions.	• The molecules can move around in any direction and are **not fixed in position**.	• The molecules move around very quickly in all directions.
• Solids have a fixed shape and volume.	• Liquids have a fixed volume but take on the shape of the container.	• Gases completely fill their container.
• Solids have a high density because their molecules are so tightly packed.	• Liquids have a medium density.	• Gases have a very low density because their molecules are so far apart.
• There are strong forces of attraction between the molecules.	• The forces of attraction between them are still **quite strong** but, again, not as strong as in solids.	• The forces of attraction between them are negligible.

Exam practice

1 (a) Explain what is meant by density. [2]
   (b) Describe briefly how you could use a measuring cylinder half filled with water to find the volume of a bracelet. In your description state what measurements you would make and what calculation you would carry out. [4]
   (c) A certain bracelet has a volume of 2.4 cm³ and a mass of 46 g. Calculate its density. [3]
   (d) The bracelet is made from a metal which is almost 100% pure. Use your answer to part (c) and the table below to find out what the metal is. [1]

Metal	Copper	Gold	Lead	Platinum
Density/g/cm³	8.9	19.3	11.3	21.5

2 (a) A rectangular concrete slab of mass 520 kg is 1.8 m long, 1.2 m wide and 0.1 m deep. Use these data to calculate the volume of this concrete slab. [4]
   (b) For bridge construction the concrete slabs must have a density of at least 2350 kg/m³. Is this particular slab dense enough to be used for bridge construction? Show clearly how you get your answer. [4]

3 The density of aluminium is 2.7 g/cm³.
   (a) Calculate the number of cm³ in 1 m³. [1]
   (b) Calculate the mass in grams of 1 m³ of aluminium. [1]
   (c) Calculate the density of aluminium in kg/m³. [2]

4 A student places an empty measuring cylinder on an electronic balance and adds different volumes of liquid. Each time she measures the volume of the liquid she also records the reading on the electronic balance. She plots her results as a graph as shown in Figure 3.4.
   (a) What is the mass of the empty measuring cylinder? [1]
   (b) The table shows four different liquids and their densities.

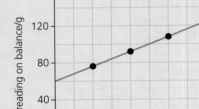

Figure 3.4

Liquid	Density/g/cm³
Petrol	0.7
Castor oil	0.9
Water	1.0
Ethanol	0.8

   Using the data from Figure 3.4 and your answer to part (a) identify the liquid the student used. [4]
   (c) The student repeats the same procedure using the same measuring cylinder but using a liquid of lower density than the first liquid. Copy the grid in Figure 3.4 and on it draw the straight line she would expect to obtain. [2]
   (d) Ice has a density of 0.9 g/cm³ and water has density of 1.0 g/cm³. What does this tell you about the spacing of the molecules in the two states? [1]

Answers online

ONLINE

4 Energy

Energy forms

REVISED

It is important to understand the difference between **energy forms** and **energy resources**. Energy forms are the different ways in which energy can appear, such as heat, light, sound, nuclear, **kinetic**, **gravitational potential** and chemical energy. Energy resources are the different ways of supplying a particular energy form. Table 4.1 summarises some of the main energy forms.

Table 4.1

Energy form	Definition	Examples of resources
Chemical	The energy stored within a substance, which is released on burning	Coal, oil, natural gas, peat (turf), wood, food
Gravitational potential	The energy a body contains as a result of its height above the ground	Stored energy in the dam (reservoir) of a hydroelectric power station
Kinetic	The energy of a moving object	Wind, waves, tides
Nuclear	The energy stored in the nucleus of an atom	Uranium, plutonium

Other common energy forms are electrical energy, magnetic energy and strain potential energy — the energy a body has when it has been stretched or squeezed out of shape and will return to its original shape when the force is removed, such as a wind-up toy.

One of the fundamental laws of physics is the **principle of conservation of energy**. This states that:

Energy can neither be created nor destroyed, but it can change its form.

In other words, energy can be changed from one form into another, but the total amount of energy does not change.

Energy resources

REVISED

Energy resources can be classified as **renewable** or **non-renewable**. Renewable resources are those that are replaced by nature in less than a human lifetime. Non-renewable resources are those that are used faster than they can be replaced by nature. We will eventually run out of non-renewable energy resources. Table 4.2 lists the common renewable energy resources and Table 4.3 lists common non-renewable resources.

energy forms are the different ways in which energy can appear, such as heat, light, sound etc.

energy resources are the different ways of supplying a particular energy form such as coal, oil, wind etc.

kinetic energy is the energy possessed by a mass due to its speed

gravitational potential energy is the energy possessed by a mass due to its height above the ground

the **principle of conservation of energy** states that energy can change from one form to another but can never be created or destroyed

renewable energy is energy that is being replaced by nature in less than a human lifetime, so we will never run out of it

non-renewable energy is energy that will eventually run out

Table 4.2 Renewable resources

Resource	Comment
Solar cells	Solar cells convert sunlight (**solar energy**) directly into electricity; they are joined together into arrays.
Hydroelectric power stations	Water behind a high dam contains **gravitational potential energy**. The water is allowed to fall from the dam through a pipe and it gains **kinetic energy** as it falls. The fast-flowing water falls on a **turbine**, which then drives a **generator**. The output from the generator is **electrical energy**.
Tidal barrages	This is created when a **dam** is built across a **river estuary**. As the tide rises and falls every 12 hours, water will flow through a gate in the dam. The moving water drives a turbine, which is made to turn a generator to produce electrical energy.
Wave machines	Waves are produced largely by the action of the wind on the surface of water. The **wave machine** floats on the surface of the water and the up-and-down motion of the water is converted to rotary motion of a turbine–generator unit to produce electricity.
Wind turbines	As the wind blows, the large blade turns and this drives a turbine. The turbine drives a generator, which produces electricity. Large numbers of turbines are often grouped together to form a **wind farm**.
Geothermal power stations	These use heat from the hot rocks deep inside the Earth. Cold water is passed down a pipe to the rocks. The rocks heat the water and the hot water is then pumped to the surface. The steam generated is used to drive a turbine–generator to produce electricity.
Biomass	The timber from fast-growing trees is harvested. The wood is dried and turned into woodchips, which are then burned in power stations to produce electricity or sold for solid fuel heating. There are many other forms of biomass. Oil from **oil-bearing seeds** can be converted into **biodiesel** for road transport. Grass can be fermented in a **bio-digester** and turned into **biogas** for heating.

Table 4.3 Non-renewable resources

Resource	Comment
Fossil fuels	These are oil, natural gas, coal, turf (peat) and lignite. They are burned in power stations to produce steam, which drives a turbine, turns a generator and so produces electricity.
Nuclear power	Large nuclei (uranium or plutonium) in a nuclear reactor are made to split into lighter nuclei (**by nuclear fission**) with the release of large amounts of kinetic energy (of sub-atomic particles). This energy is used to produce steam, which drives a turbine, turns a generator and so produces electricity.

All energy resources have unique **advantages** and **disadvantages**. The main features of the most common resources are listed in Table 4.4.

Table 4.4 Advantages and disadvantages of the main energy resources

Energy resource	Advantages	Disadvantages	Other comments
Fossil fuels — coal, oil, natural gas, lignite, turf	Relatively cheap to start up. Moderately expensive to run. Large world reserves of coal (much less for other fossil fuels).	All fossil fuels are non-renewable. All fossil fuels release carbon dioxide on burning and so contribute to global warming. Burning coal and oil also releases sulfur dioxide gas, which causes acid rain.	Coal releases the most carbon dioxide and natural gas the least per unit of electricity produced. Removing sulfur or sulfur dioxide is very expensive and adds greatly to the cost of electricity production.
Nuclear fuels — mainly uranium	Do not produce carbon dioxide. Do not emit gases that cause acid rain.	The waste products will remain dangerously radioactive for tens of thousands of years. As yet, no one has found an acceptable method to store these materials cheaply, safely and securely for such a long time. Nuclear fission fuels are non-renewable. An accident could release dangerous radioactive material, which would contaminate a wide area, leaving it unusable for decades.	Nuclear fuel is relatively cheap on world markets. Nuclear power station construction costs are much higher than fossil fuel stations, because of the need to take expensive safety precautions. Decommissioning nuclear power stations is particularly long and expensive, requiring specialist equipment and personnel.
Wind farms	A renewable energy resource. Low running costs. Conserve fossil fuels.	Wind farms are: ● unreliable ● unsightly ● very noisy ● hazardous to birds	Wind farms take up much more ground per unit of electricity produced than conventional power stations.
Waves	A renewable energy resource. Low running costs. Conserve fossil fuels.	Wave generators at sea are: ● unreliable ● unsightly ● hazardous to shipping	Many turbines are needed to produce a substantial amount of electricity.
Tides	A renewable energy resource. Low running costs. Conserve fossil fuels.	Tidal barrages are built across river estuaries and can cause: ● navigation problems for shipping ● destruction of habitats for wading birds and the mud-living organisms on which they feed	Tides (unlike wind and waves) are predictable, but they vary from day to day and month to month. This makes them unsuitable for producing a constant daily amount of electrical energy.

Now test yourself

1 Name four fossil fuels.
2 Name six different energy forms.
3 Explain what is meant by:
   (a) a renewable energy resource
   (b) a non-renewable energy resource.
4 Name four renewable energy resources and four non-renewable energy resources.
5 Why is it misleading to say that electricity is a 'clean' fuel?
6 State the principle of conservation of energy.

Answers online

Energy flow diagrams

It is common for examiners to ask about the main energy flow through various devices. The kind of question that has appeared in recent examination questions is shown in the example.

Example

Fill in the spaces below to show the main types of energy change each device is designed to bring about. (Here the answers have all been provided to show you what the examiners expect.)

Main input energy	Device	Required output energy
Sound	Microphone	Electricity
Electricity	Room heater	Heat
Chemical	Burning match	Heat

Typical mistake

A common mistake is to write that renewable resources are those that can be used over and over again. This is quite wrong. Once a unit of energy has been used, that particular unit of energy can never be used again. So *learn the correct definitions* of renewable and non-renewable resources.

The Sun

Almost all energy resources rely on the energy of the Sun. Fossil fuel resources come from the dead remains of plants and animals laid down many millions of years ago. The plants obtained their energy from the Sun by **photosynthesis**. Herbivores ate the plants, while carnivores ate the herbivores. Under the Earth's surface, these remains slowly fossilised into coal, peat, gas and oil. Other processes also rely on the Sun's energy. Hydroelectric energy depends on the water cycle, which begins when ocean water evaporates as a result of absorbing radiant energy from the Sun. Wind and waves rely on the Earth's weather, which is largely controlled by the Sun.

Only geothermal and nuclear energy do not depend directly on the energy emitted by the Sun.

Now test yourself

7 (a) Copy and complete the boxes below to show the useful energy change that takes place in a wind turbine.

input		useful output
☐	→	☐

(b) The wind is a renewable energy resource. What does this mean?

(c) Give two other examples of renewable energy resources.

8 Figure 4.1 represents a typical hydroelectric power station. Copy and complete the boxes below to show the energy changes taking place in a hydroelectric power station.

☐	→	☐	→	☐
(energy stored in the upper lake)		(energy in the moving water)		(output energy from the power station)

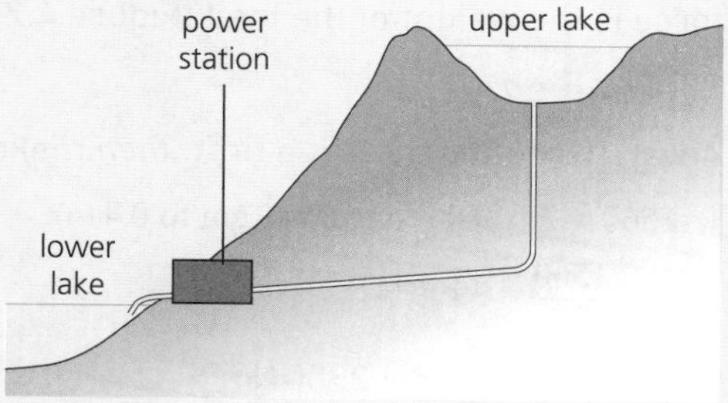

Figure 4.1

9 For each of the devices or situations shown below, construct a flow diagram to show the main energy change that is taking place. The first has been done for you.

Device/situation	Input energy form		Useful output energy form
Microphone	**Sound** energy	→	**Electrical** energy
Loudspeaker	_____energy	→	_____energy
Electric smoothing iron	_____energy	→	_____energy
Coal burning in an open fire	_____energy	→	_____energy
Weight falling towards the ground	_____energy	→	_____energy
Candle flame	_____energy	→	_____energy and _____energy
Battery-powered electric drill	_____energy → _____ energy	→	_____energy

Answers online

Work

Work is only done when a **force** causes **movement**. The amount of work done is given by the formula:

work done = force × distance moved in the direction of the force

or

$$W = F \times d$$

where work, W, is measured in newton metres (Nm) or joules (J), force, F, is measured in newtons (N), distance, d, is measured in metres (m).

> **work** is the product of the force and the distance moved in the direction of the force

> **Exam tip**
>
> This formula must be memorised.

Examples

1 How much work is done when a packing case is dragged 60 cm across the floor at a steady speed against a frictional force of 45 N? How much energy is needed?

Answer

There are two points to watch here. First, the frictional force and the direction of motion are in opposite directions. But the case moves at a steady speed, so the forward force must also be equal to the frictional force (45 N). Second, the distance moved is given in cm, so be sure to convert it to metres.

So work done = $F \times d$ = 45 × 0.6 = 27 J

2 A crane does 1560 J of useful work when it lifts a load vertically by 40 cm. Find the weight of the load.

Answer

Since the load is being lifted, the minimum upward force is the weight of the load (Figure 4.2).

So, $W = F \times d$

Always write the equation first, *then* make the substitutions.

1560 = $F \times 0.4$ (convert 40 cm to 0.4 m)

$$F = \frac{1560}{0.4} = 3900 \text{ N}$$

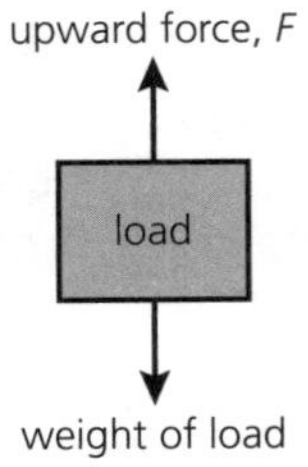

So, weight of load = 3900 N

upward force, F

load

weight of load

Figure 4.2

3 How much work is done by an electric motor pulling a 130 N load 6.5 m up the slope shown in Figure 4.3 if the constant tension in the string is 60 N?

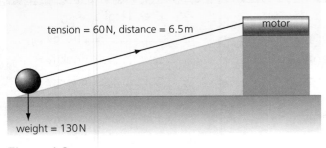

tension = 60 N, distance = 6.5 m

motor

weight = 130 N

Figure 4.3

Answer

Since the tension force and distance moved are both parallel to the slope, they are both used to find the work done. Neither the weight of the load, nor the vertical distance between the weight and the motor are used in this question.

$W = F \times d$ moved in the direction of the force

= 60 × 6.5 = 390 J

Work and energy

REVISED

Energy is the *ability* to do work. So if a machine has 100J of stored energy, this means it can do 100J of work. Similarly, work is sometimes thought of as the amount of energy transferred. Note that both work and energy are measured in joules.

> **energy** is what is required to do work

Example

A battery stores 15kJ of energy. If the battery is used to drive an electric motor, what is the *maximum* height to which it could raise a 750N load, if it was lifted vertically?

Answer

The battery stores 15kJ or 15000J, so it can do a maximum of 15000J of work.

$$W = F \times d$$

$$15000 = 750 \times d$$

$$d = \frac{15000}{750} = 20\,m$$

Power

REVISED

Power is the *rate* of doing work. This means that the power of a machine is the work it can do in a second. The formula for calculating power is therefore:

$$\text{power} = \frac{\text{work done}}{\text{time taken}}$$

or

$$P = \frac{W}{t}$$

> **power** is the rate of doing work

> **Exam tip**
>
> This formula must be memorised.

where power, P, is measured in joules per second (J/s) or watts (W)
work, W, is measured in joules (J)
time, t, is measured in seconds (s)

Examples

1   An electric motor is used to raise a load of 210N. The load rises vertically 3m in a time of 6s. Find the work done and the power of the motor.

Answer

Note: This is a two-part question. First, calculate the work done using $W = F \times d$, then find the power using $P = \frac{W}{t}$.

work done = force × distance = 210 × 3 = 630J

$$\text{power} = \frac{\text{work done}}{\text{time taken}} = \frac{630}{6} = 105\,W$$

→

2 A crane has a power of 1200W. How much work could it do in 1 hour?

Answer

In power calculations, the unit of time is the second. So first convert 1 hour to seconds:

1 hour = 60 minutes = 60 × 60 seconds = 3600 seconds

$$\text{power} = \frac{\text{work done}}{\text{time taken}}$$

$$1200 = \frac{\text{work done}}{3600}$$

work done = 1200 × 3600 = 4 320 000 J or 4.32 MJ

Measuring power

Prescribed practical P5

Measuring personal power

An experiment to measure personal power is prescribed by the specification. This means you must do it as an experiment in class. In recent years, a description of the experiment has been asked as a 6-mark QWC question.

Figure 4.4

In this experiment be sure to remember:
● what is measured
● the apparatus used to make the measurements
● the calculations required to find the output power.

To measure your personal power, you need to find out how long it takes you to do a given amount of work. First find your weight in newtons. The easy way to do this is to find your mass in kilograms using bathroom scales, and then use the fact that 1 kg has a weight of 10 N. Then you need to find the height of a staircase.

This can be done by measuring the average height of a riser (stair) using a metre ruler and multiplying by the number of risers in the staircase. Finally, you need to have someone who will time you as you run up the stairs using a stopwatch (Figure 4.4).

For reliability, the experiment should be repeated and the average power of the pupil determined.

➜

Measurements

Some typical measurements that might be obtained are:

- mass of student in kg: 45
- weight of student in N: 450
- height of risers in cm: 14.0, 13.8, 13.8, 14.0, 13.9
- average riser height in cm: 13.9
- number of risers: 30
- staircase height: $13.9 \times 30 = 417\,\text{cm} = 4.17\,\text{m}$
- time to run upstairs in s: 5.2, 5.1, 4.9, 5.0, 4.8
- average time taken in s: 5.0

Calculations

$$\text{work} = \text{force} \times \text{distance} = 450 \times 4.17 = 1876.5\,\text{J}$$

$$\text{power} = \frac{\text{work}}{\text{time}} = \frac{1876.5}{5.0} = 375\,\text{W (approx.)}$$

An alternative method involves a student of known mass, m, being timed to do, say, 100 'step-ups' on to a platform such as the first step of a flight of stairs. If the time taken is t and the height of the platform is h metres, then the power, P, of the student is given by:

$$P = 100 \times \frac{mgh}{t}$$

The apparatus used to obtain the measurements are the same as described in the stairs experiment above.

ⒻElectric motor

To measure the output power of an electric motor we measure the time, t, it takes for the motor to raise a known mass, m, through a known height, h, at a constant speed (Figure 4.5).

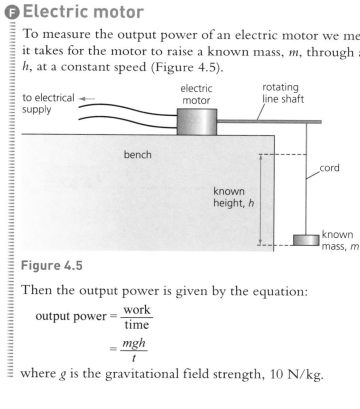

Figure 4.5

Then the output power is given by the equation:

$$\text{output power} = \frac{\text{work}}{\text{time}}$$

$$= \frac{mgh}{t}$$

where g is the gravitational field strength, 10 N/kg.

Efficiency

Efficiency is defined by the formula:

$$\text{efficiency} = \frac{\text{useful energy output}}{\text{total energy input}}$$

Alternative definitions are possible involving work and power. Learn one and stick to it.

efficiency is the ratio of the useful output work to the total energy input

Examples

1 An electric kettle is rated 2500 W. It produces 2500 J of heat energy every second. The kettle takes 160 seconds to boil some water and during this time 360 000 J of heat energy pass into the water. Find the kettle's efficiency.

Answer

useful energy output (passed into water) = 360 000 J

total energy input = 2500 × 160 = 400 000 J

$$\text{efficiency} = \frac{\text{useful energy output}}{\text{total energy input}} = \frac{360\,000}{400\,000} = 0.9$$

2 A motor lifts a load of 80 N to a height of 90 cm. In the process, 88 J of energy are wasted as heat and sound. Find the motor's efficiency.

Answer

useful energy output = work done by motor = force × distance

$$= 80 \times 0.9 = 72\,\text{J}$$

total energy input = total energy output = useful energy + wasted energy

$$= 72 + 88 = 160\,\text{J}$$

$$\text{efficiency} = \frac{\text{useful energy output}}{\text{total energy input}} = \frac{72}{160} = 0.45$$

Typical mistake

A common mistake is to have the input on the top line and the output on the bottom. As efficiency is a ratio, it has no units. And since *some* energy is wasted in *every* physical process, the efficiency of a machine is *always less than 1*.

Exam tip

Avoid the use of percentages in efficiency calculations. For example, giving the efficiency of the motor in the last example as 45% will get full marks. But leaving out the % sign and writing only 45 will cost you at least 1 mark.

Gravitational potential energy

You are expected to remember that 1 kg has a weight of 10 N on Earth.

This is just another way of saying that the gravitational field strength, g, on Earth is 10 N/kg. So the weight of an object, W, is given by:

$$W = mg$$

When any object with mass is lifted, work is done on it against the force of **gravity**. The greater the mass of the object and the higher it is lifted, the more work has to be done. The work that is done is only possible because some energy has been used. This energy is stored in the object as **gravitational potential energy (GPE)**.

GPE = work done in raising a load (m) against the force of gravity (g) through a height (h)

$$\text{GPE} = mgh$$

where m is the mass in kg
g is the gravitational field strength in N/kg
h is the vertical height in m.

Exam tip

This formula must be memorised.

Answers at **www.hoddereducation.co.uk/myrevisionnotesdownloads**

Examples

1   Find the gravitational potential energy of a mass of 600 g when raised to a height of 250 cm. Take $g = 10$ N/kg.

Answer

First note that 600 g is 0.6 kg and 250 cm is 2.5 m.

$GPE = mgh = 0.6 \times 10 \times 2.5 = 15$ J

2   How much heat and sound energy is produced when a mass of 1.2 kg falls to the ground from a height of 5 m? Take $g = 10$ N/kg.

Answer

Heat and sound energy produced = original GPE = $mgh = 1.2 \times 10 \times 5$
$= 60$ J

3   A book of mass 500 g has a gravitational potential energy of 3.2 J when at a height of 4 m above the surface of the Moon. Find the gravitational field strength on the Moon.

Answer

$GPE = mgh = 3.2$

$3.2 = 0.5 \times g \times 4 = 2g$

$g = \dfrac{3.2}{2} = 1.6$ N/kg

Kinetic energy

The **kinetic energy (KE)** of an object is the energy it has because it is moving. It can be shown that an object's kinetic energy is given by the formula:

$KE = \frac{1}{2} mv^2$

where m is the mass in kg
v is the speed of the object in m/s

> **Exam tip**
>
> If you find it hard to rearrange the kinetic energy formula to find speed, you might remember
> $v = \sqrt{\dfrac{2 \times KE}{m}}$

Examples

1   A car of mass 800 kg is travelling at 15 m/s. Find its kinetic energy.

Answer

$KE = \frac{1}{2} mv^2 = \frac{1}{2} \times 800 \times 15^2 = 0.5 \times 800 \times 225 = 90\,000$ J

2   A bullet has a mass of 20 g and is travelling at 300 m/s. Find its kinetic energy.

Answer

We must first change the bullet's mass from g to kg by dividing by 1000.

$KE = \frac{1}{2} mv^2 = \frac{1}{2} \times \left(\dfrac{20}{1000}\right) \times 300^2 = 0.5 \times 0.02 \times 90\,000 = 900$ J

3   The input power of a small hydroelectric power station is 1 MW. If 18 000 000 kg of water flows past the turbines every hour, find the average speed of the water.

→

Answer

1 hour = 60 × 60 seconds = 3600 seconds
mass of water flowing every second = 18 000 000/3600 = 5000 kg/s

Since a 1 MW power station produces 1 000 000 J of electrical energy per second, the minimum KE of the water passing every second is 1 000 000 J.

So KE = ½ mv^2

$$1\,000\,000 = \tfrac{1}{2} \times 5000 \times v^2$$
$$v^2 = \frac{1\,000\,000}{2500}$$
$$v^2 = 400$$
$$v = \sqrt{(400)} = 20\,\text{m/s}$$

Now test yourself

10 Explain why a waiter holding a heavy tray at rest for 5 minutes is doing no work.
11 Write down the equations for work, power and efficiency.
12 Explain the difference between gravitational potential energy and kinetic energy and write down the equation for each.
13 Explain why no machine can have an efficiency greater than 1.
14 Explain why efficiency has no unit.
15 According to your specification it takes approximately 1 J to lift an apple vertically 1 m. Use this information to calculate the typical mass of an apple.
16 (a) How much work is done by a tractor when it lifts a load of 8000 N to a height of 1.8 m?
   (b) The output power of the tractor is 5.2 kW. How long does it takes to do 26 000 J of work?
   (c) The efficiency of the tractor is 0.26 (or 26%). If the output power of the tractor is 5.2 kW, calculate the input power.
17 Stephen weighs 550 N. How much work does he do in climbing up to a diving board which is 3.0 m high?
18 A basketball player throws a ball vertically up into the air. Place a tick (✓) in the appropriate column to show what happens to each quantity as the ball rises. Ignore the effects of friction.

Quantity	Increases	Decreases	Remains constant
Speed of ball			
Potential energy of ball			
Total energy of ball			
Kinetic energy of ball			

Answers online

Exam practice

1 Competitors in the World's Strongest Man competition must throw a cement block of mass 100 kg over a wall 5.5 m high. How much work is done if the block just clears the top of the wall? [4]

2 A gardener pushes a lawn mower 200 m with a force of 60 N. The work takes her 4 minutes.
   (a) How much work does she do altogether? [4]
   (b) Calculate her average power. [4]

3 The electrical energy used by a boiler is 1050 kJ. The useful output energy is 840 kJ.
   (a) Calculate the efficiency of the boiler. [3]
   (b) Suggest what might have become of the energy wasted by the boiler. [1]

4 A car engine has an efficiency of 0.28. How much input chemical energy must be supplied if the total output useful energy is 140 000 kJ? [4]

5 The power of the electric motor in a lift is 3600 W. How much electrical energy is converted into other energy forms in 3 minutes if the lift has been rising continuously? [4]

6 A barrel of weight 1000 N is pushed up a ramp. The barrel rises vertically 40 cm when it is pushed 1 m along the ramp.
   (a) Calculate how much useful work is done when the barrel is pushed 1 m along the ramp. [4]
   (b) To push the barrel 1 m along the ramp requires 1200 J of energy. Calculate the efficiency of the ramp. [3]

7 A communications satellite of mass 120 kg orbits the Earth at a speed of 3000 m/s. Calculate its kinetic energy. [3]

8 On planet X an object of mass 2 kg is raised 10 m above the surface. At that height the object has a gravitational potential energy of 176 J. Details of three planets are given in the table. Which one of these three planets is most likely to be planet X? [4]

Planet's name	Mercury	Venus	Jupiter
Gravitational field strength, g, in N/kg	3.7	8.8	26.4

9 A ball of mass 2 kg falls from rest at a height of 5 m above the ground. Copy the table below and complete it to show the gravitational potential energy, the kinetic energy, speed and the total energy of the falling ball at different heights above the surface. [7]

Height above ground/m	Gravitational potential energy/J	Kinetic energy/J	Total energy/J	Speed/m/s
5.0		0	100	0
4.0				4.47
	64			
1.8		64		
0.0	0			

10 A bouncing ball of mass 200 g leaves the ground with a kinetic energy of 10 J.
   (a) If the ball rises vertically, calculate the maximum height it is likely to reach. [3]
   (b) In practice, the ball rarely reaches the maximum height. Explain why this is so. [1]

Answers online

5 Heat transfer

(F) There are three main methods of heat transfer: conduction, convection and radiation.

Conduction

REVISED

Conduction in metals

The best conductors are metals. Figure 5.1 shows an experiment to demonstrate which of four materials is the best heat conductor.

To make it a **fair test**, all the rods must be:
- the *same* length
- the *same* area of cross-section
- placed at the *same* point in the Bunsen flame

The drawing pins must also be identical and attached to the rods with the same mass of candle wax.

Heat conducts along all of the rods, but the pins fall off at different times. The pin attached to the copper rod falls first, shortly followed by the pin attached to the aluminium rod and then the pin attached to the steel rod. Only after many minutes does the pin attached to the glass rod fall.

The best conductors (in order) are:
1 copper
2 aluminium
3 steel

Glass is a *very poor* heat conductor.

Poor conductors are called **insulators**.

> **conduction** is the transfer of heat from molecule to adjacent molecule by vibrations

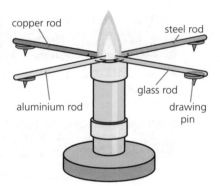

Figure 5.1 This apparatus can be used to demonstrate heat conduction in different materials

Metals are good conductors because:
- they have **free electrons** that can move freely throughout the solid structure
- free electrons absorb heat from the Bunsen flame, causing them to move *much faster*
- the fast, free electrons collide with the metal's atoms as they move through the solid structure
- a free electron gives up a little kinetic energy in each collision
- these collisions cause the atoms to vibrate faster than before.

Non-metals (insulators) such as glass:
- have no free electrons
- so glass atoms near the flame absorb heat directly
- this makes the atoms in the end of the rod vibrate *faster* than before
- these vibrations pass from atom to atom through the solid structure
- the process is much slower than electron conduction in metals.

1 Container ships are used to carry fruit and vegetables all over the world. The hold of the ship has two metal walls with an insulator in-between.
   (a) What is the purpose of the insulating material?
   (b) Give the name of a suitable insulator for this purpose.
   (c) What makes this insulator effective?

Answers online

Convection

REVISED

Convection occurs only in liquids and gases.

Convection occurs when the fastest-moving particles in a hot region of a gas or liquid move to a cool region, taking their heat energy with them. Convection is explained by changes in the density of the material.

> **convection** is the transfer of heat through a liquid or gas by the motion of molecules themselves

Convection in a gas

Convection in air can be demonstrated by the glass chimney experiment shown in Figure 5.2. The smoking straw is held over each chimney in turn.

When the smoking straw is over the candle flame, the smoke *rises*.

When the straw is held over the other chimney, the smoke *falls*.

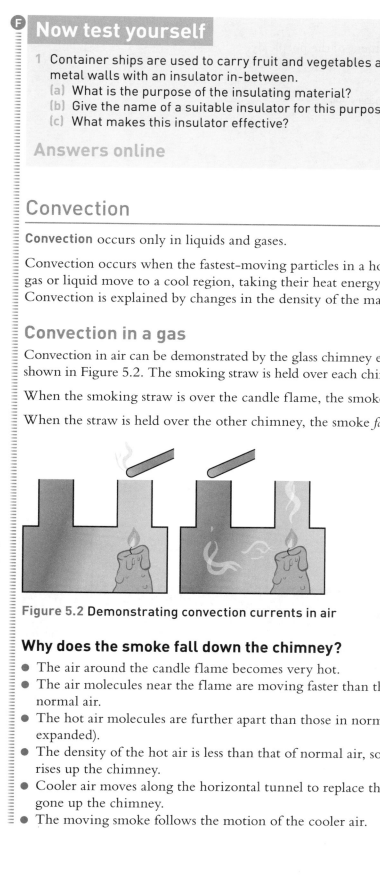

Figure 5.2 Demonstrating convection currents in air

Why does the smoke fall down the chimney?

- The air around the candle flame becomes very hot.
- The air molecules near the flame are moving faster than those in normal air.
- The hot air molecules are further apart than those in normal air (air has expanded).
- The density of the hot air is less than that of normal air, so the hot air rises up the chimney.
- Cooler air moves along the horizontal tunnel to replace the air that has gone up the chimney.
- The moving smoke follows the motion of the cooler air.

Convection in a liquid

Figure 5.3 shows convection in water. The movement of the purple dye in the water shows the convection current.

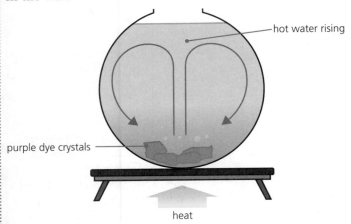

Figure 5.3 Demonstrating convection currents in a liquid

What causes convection currents in water?

As the water at the bottom of the flask warms up, the molecules gain kinetic energy. This extra energy causes the following to happen:
- The molecules vibrate with greater amplitude.
- The warm water therefore expands.
- The density of the warm water is less than that of the cold water.
- The warm water rises.
- Cooler water flows downwards to replace the upward-moving warmer water.
- Cool water at the top falls as it is replaced by warm water.

> **Exam tip**
>
> When explaining convection you must not write 'heat rises'.

Reducing heat loss from your home

REVISED

Table 5.1 Types of home insulation

Device	Payback time (using money saved on heating bills)	How losses are reduced
Cavity wall insulation	3 years	• The cavity between the outside walls is filled with fibreglass, mineral wool or foam. • Mineral wool and foam both trap air in tiny pockets. • Trapped air reduces heat loss through walls by convection and conduction.
Loft (attic) insulation	1.5 years	• Fibreglass or mineral wool fibres trap air. • Trapped air reduces heat loss through the roof by convection and conduction.
Double glazing	40 years	• Thick glass is used to reduce heat loss through windows by conduction. Air trapped between the glass panes reduces heat loss by convection and conduction.
Thick curtains and carpets	Variable (depends on quality)	• Trapped air reduces heat loss through windows and floors by convection and conduction.

Ⓕ Heat is lost through the roof, walls, windows and floor of a home. Different materials and devices have been designed to reduce this heat loss. Different types of **home insulation** are summarised in Table 5.1.

> **home insulation** is the use of insulating materials to reduce the loss of heat energy from a home

Now test yourself

Use the information in Table 5.1 to help you to answer the following.
2 (a) Explain what is meant by saying that the payback time for double glazing is 40 years.
  (b) Suggest another reason why people might have their windows double glazed. (Something other than they look good and last a long time.)
3 Loft insulation saves a householder £90 each year in reduced fuel bills. What is the cost of the loft insulation?

Answers online

Radiation

Radiation is the method of heat transfer that takes place without the need for any particles. So radiation can take place in a vacuum. The heat energy is transferred as **infrared waves**.

> **radiation** is the transfer of energy by electromagnetic waves, usually infrared waves

You need to know what types of materials are best at **emitting**, **absorbing** and **reflecting** radiant heat.

All objects radiate energy (emit radiant heat). The *hotter* an object is, the *more* radiation it emits. All objects also absorb radiant heat. If an object is hotter than its surroundings, it emits more radiant heat than it absorbs, so its temperature falls. If an object is *cooler* than its surroundings, it absorbs *more* radiant heat than it emits, so its temperature rises.

Giving out radiation (radiation emission)

Figure 5.4 shows an experiment in which a thick piece of copper is covered with gloss (shiny) white paint on one side and matt (non-shiny) black paint on the other. The copper has been heated with a Bunsen burner until it is very, very hot.

A hand held about 30 cm from the matt black side feels very much hotter than when it faces the gloss white side. This is because the matt black surface is the better emitter of radiant heat.

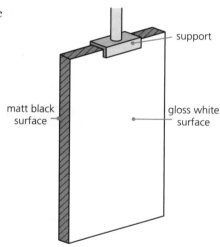

Figure 5.4 Investigating radiation emission

Absorbing radiation

Figure 5.5 shows the apparatus used to demonstrate which surfaces are the best absorbers of radiation and which are the best reflectors. Identical corks are attached with the same amount of candle wax to identical aluminium plates. Both the aluminium plates are the same distance from the Bunsen flame, so each receives the same amount of radiant heat. However, the cork attached to the dull black plate falls off first because that plate is the better at absorbing radiant heat. The polished gloss white plate is the better reflector of radiation.

dull black

gloss white

cork

cork

Figure 5.5 Investigating radiation absorption

Rules to remember

- Radiant heat falling on a surface is partly absorbed and partly reflected.
- Matt black surfaces are good absorbers (and poor reflectors) of radiation.
- Gloss white surfaces are poor absorbers (and good reflectors) of radiation.

Table 5.2 summarises radiation emission and absorption.

Table 5.2 Radiation summary

Matt black surfaces	Gloss white surfaces
Good emitters and absorbers of radiation	Poor emitters and absorbers of radiation
Poor reflectors of radiation	Good reflectors of radiation

Applications of heat transfer

REVISED

Vacuum flasks

The **vacuum flask** was designed by James Dewar in order to keep liquids cold. But the flask works equally well as a way of keeping liquids hot. Today it is commonly used as a picnic flask to keep tea, coffee or soup hot. How does it work?

- The flask is made of a double-walled glass bottle (or stainless steel in unbreakable flasks). There is a vacuum between the two walls (Figure 5.6). The vacuum stops all heat transfer by conduction or convection through the sides.
- The glass walls facing the vacuum are silvered. Their shiny surfaces reduce heat transfer by radiation to a minimum.
- The stopper is made of plastic and is often filled with cork or foam to reduce heat transfer through it by conduction.
- The outer, sponge-packed plastic case protects the inner, fragile flask against physical damage.

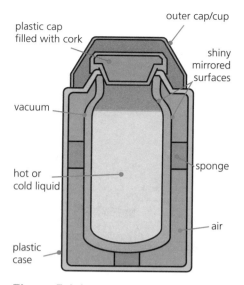

Figure 5.6 **A cross-section through a vacuum flask**

labels: plastic cap filled with cork, outer cap/cup, shiny mirrored surfaces, vacuum, hot or cold liquid, sponge, air, plastic case

Solar panels

A solar panel (Figure 5.7) absorbs sunlight and uses the energy to heat water in the following process:

● The sunlight passes through a glass window and falls on to a blackened metal sheet.
● The metal is in a draught-proof enclosure to minimise heat loss by convection.
● The blackened metal absorbs almost all of the energy in the sunlight and its temperature often rises to over 100°C.
● The heat stored in the blackened metal is then transferred to water flowing in a nearby pipe.

Solar panels are ideal for use where large volumes of hot water are needed, such as swimming pools and hospital laundries.

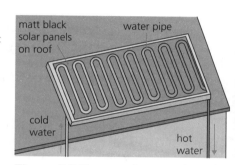

Figure 5.7 **How a solar panel works**

labels: matt black solar panels on roof, water pipe, cold water, hot water

Now test yourself

TESTED

4 As food is cooled in a fridge, heat energy is transferred to a coolant. The coolant, usually a liquid with a low boiling point, passes through pipes at the back of the fridge (Figure 5.8). The pipes are usually painted black and have thin metal 'fins' attached.

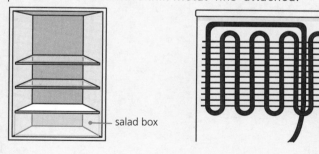

label: salad box

Figure 5.8

(a) Why are the pipes painted black?
(b) Why are the pipes mounted on thin metal fins?
5 (a) Which part of an oven is hottest, the top or the bottom?
(b) Why is this so?
(c) What is the purpose of the fan in a fan-assisted oven?
6 Suggest a reason why the roof of a house in a very hot country is often painted white.

Answers online

Exam practice

1 (a) Copy and complete the table below to indicate the main method of heat transfer in each substance. [3]

Substance	Method of heat transfer
Water	
Copper	
Glass	

(b) Water is heated in a **glass** saucepan as shown in Figure 5.9. Describe fully, in terms of particles, how heat is transferred through the glass from the hotplate to the water. [2]

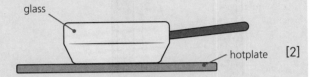

Figure 5.9

2 (a) A beaker of water is heated using a Bunsen burner as shown in Figure 5.10. Copy the diagram. Draw an arrow at A and another arrow at B to show the direction of water movement. [2]

(b) Two conical flasks contain the same amount of water at the same temperature. They are placed at equal distances from a radiant heater as shown in Figure 5.11. Why does the temperature of the water in the flask with the dull black paint rise more than in the flask with the shiny silver paint? [2]

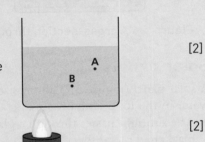

Figure 5.10

[2]

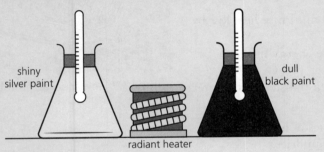

Figure 5.11

3 Figure 5.12 shows two spoons being held in ice water. One is made of steel and the other is made of plastic.

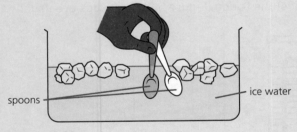

Figure 5.12

(a) Which spoon will feel colder? [1]

(b) Explain fully the reason for your answer to (a). [1]

Answers online

ONLINE

6 Atomic and nuclear physics

The structure of atoms
REVISED

All matter is made up of **atoms**, but what are atoms made of? Experiments carried out by Thomson and Rutherford in the early part of the twentieth century led physicists to believe that atoms themselves had a structure.

(H) When a current goes through a metal wire, the wire gets hot. If the wire is hot enough, its atoms emit negatively charged particles, which we now call **electrons**.

How are the electrons arranged in atoms? One of the earliest models was the '**plum-pudding**' or '**currant bun**' model in which electrons were dotted throughout the atom like currants in a bun (Figure 6.1). The positive **charge** was thought to spread throughout the atom like the dough of the bun.

> an **atom** is the smallest part of an element that can exist on its own
>
> **charge** is the property of matter that causes electrical effects

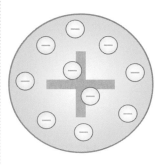

Figure 6.1 The 'plum-pudding' model of the atom

Rutherford's nuclear model
REVISED

In 1911, partly to test Thomson's theory, Rutherford suggested that the recently discovered positively charged alpha particles (see page 47) might be fired at a thin gold foil. The observations from the experiment were:
- most of the alpha particles went straight through the foil
- a few alpha particles were slightly deflected
- some were deflected through very large angles
- 1 in 8000 were 'back-scattered'.

These observations could not be explained by the 'plum-pudding' model. But what really shocked the Rutherford team was that some alpha particles were deflected through very large angles and a few even came straight back at them. Rutherford then realised that there had to be something 'hard' inside the atom to cause this strange 'back-scattering'. He called it the atomic **nucleus**.

> **nucleus** is the particle that exists at the centre of every atom and generally contains protons and neutrons

Rutherford's deductions

1 The majority of the alpha particles passed straight through the metal foil because they did not come close enough to any repulsive positive charge at all. The atom was really just empty space.

H 2 All the positive charge and most of the mass of an atom formed a dense core or nucleus.

3 The negative charge consisted of a 'cloud of electrons' surrounding the positive nucleus.

4 Only when a positive alpha particle approached sufficiently close to a nucleus was it repelled strongly enough to rebound at large angles.

5 The small size of the nucleus explained the small number that were 'back-scattered' in this way.

In 1913, in order to explain how certain elements gave out light, Niels Bohr suggested that the electrons orbited the nucleus in circular paths, like planets orbiting the Sun.

The model accepted by physicists today is called the **Rutherford–Bohr model**.

It was not until 1933 that James Chadwick proved there were two different types of particle in the nucleus — uncharged neutrons as well as positively charged protons.

The relative masses and charges of the particles that make up the atom are given in Table 6.1.

Table 6.1

Particle	Location	Relative mass	Relative charge
Proton	Within the nucleus	1	+1
Neutron	Within the nucleus	1	0
Electron	Orbiting the nucleus	1/1840	−1

Atomic number and mass number

The number of protons in a nucleus is called the **atomic number** and is given the symbol Z. The atomic number also tells you the number of electrons in the neutral atom.

The **mass number** (or nucleon number) is the sum of the number of protons and the number of neutrons. Mass number is given the symbol A.
- Atomic number, Z = number of protons
- Mass number, A = number of protons + number of neutrons = number of nucleons

Every nucleus can therefore be written in the form: $^{A}_{Z}X$ where X is the chemical symbol, A is the mass number and Z is the atomic number.

For example, the element uranium has the chemical symbol U. All uranium nuclei have 92 protons in the nucleus. One form of uranium, called uranium–235, has a mass number of 235. This means it has 92 protons and 143 neutrons (235 − 92 = 143). A uranium nucleus is given the symbol:

> **Typical mistake**
>
> The atomic number is given the symbol Z (not A). The mass number is given the symbol A (not M).

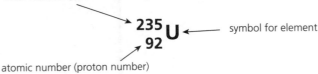

mass number (nucleon number)

$$^{235}_{92}U$$ symbol for element

atomic number (proton number)

Answers at **www.hoddereducation.co.uk/myrevisionnotesdownloads**

It is important to realise that this is the symbol for the **nucleus** of the atom. Orbiting electrons are completely ignored.

You will notice that the top number gives the mass of the nucleus and the bottom number gives its charge. This same system can also be used to describe protons, neutrons and electrons:

- proton, ^1_1p
- neutron, ^1_0n
- electron, $^0_{-1}e$

Isotopes

Not all the atoms of the same element have the same mass, but they all have the same number of protons. Physicists call atoms with the same number of protons but a different number of neutrons, **isotopes**.

Isotopes are atoms of the same element that have the same atomic number but different mass number. The main isotopes of helium, for example, are: ^3_2He and ^4_2He.

Now test yourself   TESTED ☐

1 An atom contains *electrons*, *protons* and *neutrons*. Which of these particles:
   (a) are outside the nucleus
   (b) are uncharged
   (c) have a negative charge
   (d) are nucleons
   (e) are the lightest?
2 The element sodium has the chemical symbol Na. In a particular sodium isotope there are 12 neutrons. In a neutral sodium atom there are 11 orbiting electrons. Write down the symbol for the nucleus of this isotope.
3 In what way are the nuclei of isotopes the same? In what way are they different?

Answers online

Nuclear radiation   REVISED ☐

French scientist Henri Becquerel discovered that certain rocks containing uranium gave out strange radiation that could penetrate paper and fog photographic film. He called the effect **radioactivity**. Three separate types of radiation, called **alpha** (α), **beta** (β) and **gamma** (γ) radiation were identified.

The atoms that emit these radiations are said to be **radioactive**: The particles and waves are referred to as **nuclear radiation**. The disintegration is called **radioactive decay**.

---

radioactivity is the process that occurs when alpha particles, beta particles or gamma waves are emitted from an unstable nucleus

alpha particles are the helium nuclei containing two protons and two neutrons which are emitted by radioactive nuclei

beta particles are fast-moving electrons emitted by radioactive nuclei

gamma radiation is a high-energy electromagnetic wave emitted by radioactive nuclei

---

Ionising radiation

Ions are charged atoms (or molecules). Atoms become ions when they lose (or gain) electrons (Figure 6.2).

Nuclear radiation can become dangerous by removing electrons from atoms in its path, so it has an ionising effect.

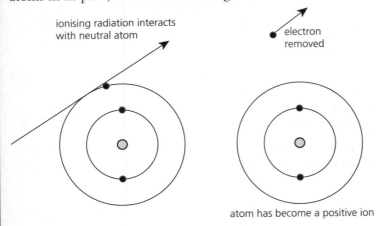

ionising radiation interacts with neutral atom

electron removed

atom has become a positive ion

Figure 6.2

When this happens with molecules in living cells, the genetic material of a cell is damaged and the cell may become cancerous. Other forms of ionising radiation include ultraviolet and X-rays.

Alpha radiation is not as dangerous if the radioactive source is *outside* the body, because it cannot pass through the skin and is unlikely to reach cells inside the body. Beta and gamma radiation can penetrate the skin and cause damage to cells. Alpha radiation will damage cells if the radioactive source has been breathed in or swallowed.

A summary of the properties of radiation is given in Table 6.2.

Table 6.2 summary of the nature and properties of nuclear radiations

Type of radiation	Alpha particle (α)	Beta particle (β)	Gamma rays (γ)
Nature	Each particle is two protons plus two neutrons; it is a nucleus of helium-4	A fast electron	Very high-energy electromagnetic waves
Source	A radioactive nucleus	A radioactive nucleus	A radioactive nucleus
Relative charge (compared with charge on proton)	+2	−1	0
Mass	High compared to beta	Low	0
Speed	Up to 0.1 × speed of light	Around 10% of the speed of light	Speed of light
Ionising effect	Strong	Weak	Very weak
Penetrating effect	Very low penetrating power, stopped by a few cm of air or thin tissue paper	Penetrating, but stopped by a few millimetres of aluminium or other metal	Very penetrating: never completely stopped, although lead and thick concrete will reduce intensity

Dangers of radiation

REVISED

Most radioactive **background activity** comes from natural sources such as cosmic rays from space, rocks and soil, some of which contain radioactive elements such as radon gas. Living things and plants absorb radioactive materials from the soil, which are then passed along the food chain.

Human behaviour also adds to the background activity that we are exposed to through medical X-rays, radioactive waste from nuclear power plants and the radioactive fallout from nuclear weapons testing.

Radioactive material is found naturally all around us and inside our bodies. Traces of radioactive elements, for example potassium, can be found in our food. Certain rocks contain uranium, all the isotopes of which are radioactive, and this decays giving radon, a radioactive gas. There is also radiation reaching Earth from outer space, referred to as cosmic rays.

All these natural sources are known together as **background radiation.**

Protection when handling radioactive material

We can minimise the risk to those using radioactive materials by:
- wearing protective clothing
- keeping the source as far away as possible by using tongs
- working quickly to keep exposure to the source to as short a time as possible
- keeping radioactive materials in lead–lined containers.

Students under the age of 16 are forbidden to handle radioactive sources.

ⓗ Nuclear disintegration equations

REVISED

Symbol equations can be written to represent alpha and beta decay. The alpha particle can be written as $_2^4\alpha$ or $_2^4$He and the beta particle as $_{-1}^0\beta$ or $_{-1}^0$e.

For example, the symbol equation for the alpha decay of uranium-238 is:

$$_{92}^{238}U \rightarrow {}_{90}^{234}Th + {}_2^4He \text{ (or } \alpha)$$

The symbol equation for the beta decay of carbon-14 is:

$$_6^{14}C \rightarrow {}_7^{14}N + {}_{-1}^0e \text{ (or } \beta)$$

H When writing symbol equations it is important to remember the following:

- The sum of the mass numbers (at the top) on the left-hand side of the equation must equal the sum of the mass numbers on the right-hand side.
- The sum of the atomic numbers (at the bottom) on the left-hand side of the equation must equal the sum of the atomic numbers on the right-hand side.

Exam tip

For all particles, the top number shows the mass and the bottom number gives the charge.

Example

Radium-226 decays to polonium-222. Radium (Ra) has atomic number 86 and polonium (Po) has atomic number 84. Which type of decay occurs?

Answer

$$^{226}_{86}Ra \rightarrow \, ^{222}_{84}Po + \, ^{a}_{b}X$$

Balancing mass numbers: $226 = 222 + a$

$a = 4$

Balancing atomic numbers: $86 = 84 + b$

$b = 2$

The particle with a mass number of 4 and an atomic number of 2 is helium and so X is an alpha particle.

Alpha decay is shown by the equation:

$$^{A}_{Z}X \rightarrow \, ^{A-4}_{Z-2}Y + \, ^{4}_{2}He \text{ (or } ^{4}_{2}\alpha)$$

If the mass number of the parent nucleus does not change and the atomic number of the daughter nucleus increases by 1, then the reaction must be beta decay. **Beta decay** is exemplified by:

$$^{A}_{Z}X \rightarrow \, ^{A}_{Z+1}Y + \, ^{0}_{-1}e \text{ (or } ^{0}_{-1}\beta)$$

To take a specific case, radium-228 decays to actinium-228 by emitting a β-particle:

$$^{228}_{88}Ra \rightarrow \, ^{228}_{89}Ac + \, ^{0}_{-1}\beta$$

In **gamma decay**, the excited parent nucleus relaxes by emitting gamma ray(s). There is no change in the nature of the nucleus, so the mass number and the atomic number stay the same. The γ-radiation is usually emitted at the same time as the α- and β-particle emissions and represents the excess energy of the daughter nucleus as it settles down into a more stable condition.

An asterisk indicates that a nucleus is in an excited state with excess energy. The process of getting rid of that energy by gamma-ray emission is called relaxation.

$$^{A}_{Z}X^* \rightarrow \, ^{A}_{Z}X + \gamma$$

(H) **Now test yourself**

6 Copy and complete the table below.

Radiation	Atomic number (Z)	Mass number (A)
α-emission	Decreases by 2	
β-emission		Unchanged
γ-emission		

7 Complete the following equations for alpha decay and beta decay.

(a) $^{238}_{92}U \rightarrow {}^{?}_{?}Th + {}^{4}_{2}He$

(b) $^{14}_{6}C \rightarrow {}^{?}_{?}N + {}^{0}_{-1}e$

8 Work out the type of decay in each of the following examples:

(a) bismuth-213 to polonium-213

(b) radium-226 to radon-222

(c) francium-221 to actinium-217.

Answers online

Radioactive decay

Radioactive decay is **random** and **spontaneous**. Random means that we cannot predict when a particular nucleus will disintegrate. Spontaneous means that the rate of decay is unaffected by any physical changes such as temperature, pressure or chemical changes. However, some types of nuclei are more unstable (Figure 6.3) than others and decay at a faster rate.

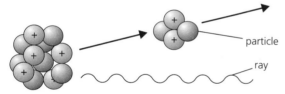

particle

ray

Figure 6.3 An unstable nucleus emitting a particle and a ray

Rate of decay and half-life

The **half-life** of a radioactive material is the time taken for its activity to fall to half its original value (Figure 6.4).

Calculations involving half-life are generally best solved using a table, as in Exam practice question 1 at the end of this chapter.

half-life is the time taken for the activity of a radioactive material to fall by half

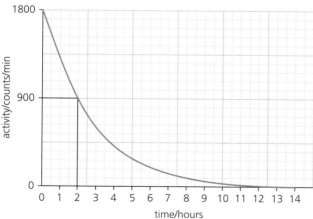

Figure 6.4 The radioactive decay curve for a substance with a half-life of 2 hours

Each isotope has a specific and constant half-life. Some half-lives are very short — a matter of seconds or even a fraction of a second — and others can be thousands of years. Table 6.3 gives the half-lives of some common radioactive isotopes.

Table 6.3

Isotope	Half-life
Uranium-238	4 500 000 000 years
Carbon-14	5730 years
Phosphorus-30	2.5 minutes
Oxygen-15	2.06 minutes
Barium-144	114 seconds
Polonium-216	0.145 seconds

The unit for radioactivity is the becquerel (Bq). 1 Bq = 1 disintegration per second.

Now test yourself

9 Calculate the half-lives of the following samples:
 (a) A sample of iodine-123 whose activity falls from 1000 Bq to 250 Bq in 14.4 hours.
 (b) A sample of technetium-99 whose activity falls from 200 Bq to 25 Bq in 18 hours.
 (c) A sample of strontium-90 whose activity falls from 500 Bq to 62.5 Bq in 86.4 years.
10 Calculate how long it would take for the following to decay to an activity of 1 Bq.
 (a) A sample of cobalt-60 (half-life = 5.27 years) whose original activity is 64 Bq.
 (b) A sample of iodine-131 (half-life = 8 days) whose original activity is 128 Bq.
 (c) A sample of polonium-210 (half-life = 138 days) whose original activity is 32 Bq.

Answers online

Uses of radiation

In medicine

Gamma radiation from the cobalt-60 isotope can be used to treat tumours.

Different radioisotopes are used to monitor the function of organs by injecting a small amount into the bloodstream and detecting the emitted radiation. The tracers used in this case must have a short half-life.

Iodine-131 is used in investigations of the thyroid gland.

Surgical instruments and hospital dressings can be sterilised by exposure to gamma radiation. The source should have a very long half-life so that it does not need to be replaced on a regular basis.

In agriculture

Gamma radiation can be used to treat fresh food. By killing bacteria on the food, the radiation helps the food to have a longer shelf life. The use is controversial, however, as many people are worried about eating food exposed to radiation. Ideally the radioisotope used in a food processing plant should have a very long half-life so that it is a long time before it needs to be replaced.

The ease with which a plant absorbs a fertiliser can be found by putting a small amount of radioactive isotope in the fertiliser. If parts of the plants are then checked for radioactivity you can tell how much fertiliser has been taken up by the plant.

In industry

Beta radiation can be used to monitor the thickness of a sheet of paper or aluminium (Figure 6.5). An emitter is placed on one side of the sheet and a detector on the other. As the sheet moves past, the activity detected will be the same as long as the thickness remains unchanged.

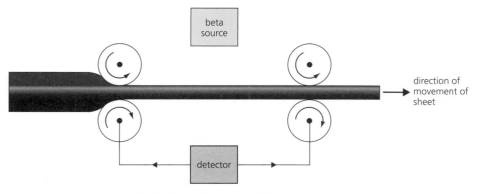

Figure 6.5 A long half-life beta source is used to control the thickness of an aluminium sheet

Radioactive tracers

A suitable radioactive isotope can be used to provide information about fluid movement and mixing to monitor, for example, leaks in underground pipes (Figure 6.6). The tracer is added to the fluid in the pipe and builds up in the ground if there is a leak. The radiation needs to penetrate many centimetres of soil to reach the detectors — this means that it must be a gamma emitter. Only gamma rays have sufficient penetrating power. But to avoid dangerous radioactive materials being in the ground for a long time the source should have a short half-life.

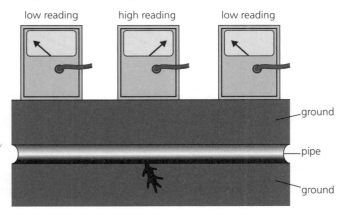

Figure 6.6 Radioactive tracers locating leaks in pipes

Practical work with radioactive materials

Students under the age of 16 are forbidden to handle radioactive sources.

The most common type of radiation detector is the Geiger–Müller tube (GM tube) connected to a counter or ratemeter. It is not necessary to know how a GM tube works, but it is important to know how it could be used to do practical work on radiation.

Background radiation

Background activity is that which is detected when no known radioactive sources are present. Fortunately, the background count in Northern Ireland does not present a serious health risk.

The background count must always be *subtracted* from any other count when measuring the activity from a specific source.

Safety precautions when using closed radioactive sources in schools

- Always store the sources in a lead-lined box, under lock and key, when not required for experimental use.
- Always handle sources using tongs, holding the source at arm's length and pointing it away from any bystander.
- Wear protective clothing.
- Always work quickly and methodically with sources to minimise the time of exposure and hence the dose to the user.

Measuring the approximate range of radiation

Alpha

- Place a GM tube on a wooden cradle and connect it to a ratemeter.
- Hold an alpha source directly in front of the window of the tube and slowly increase the distance between the source and the tube. At about 3 cm (depending on the source used) the ratemeter reading falls dramatically to that of background radiation.
- Place a thin piece of paper in contact with the window of the GM tube. Bring the alpha source up to the paper so that the casing of the source touches it. The reading on the ratemeter is now equal to the background count showing that the alpha particles are unable to penetrate the paper.

Beta

- Place a 1 mm thick piece of aluminium in contact with the window of the GM tube.
- Bring the beta source up to the aluminium so that the casing of the source touches it.
- The reading on the ratemeter is observed to be significantly above the background count, showing that some beta particles have penetrated the aluminium.
- Repeat the process with 2 mm, 3 mm etc. thick pieces of aluminium. At about 5 mm there is a significant reduction in the count rate on the ratemeter, indicating the approximate range of beta particles in aluminium.

Gamma

If the beta particle experiment is repeated with a gamma source, there is practically no reduction in the count rate for a 5 mm thick piece of aluminium. If the aluminium sheets are replaced with lead, it will be found that even school sources will give gamma radiation that can easily penetrate several centimetres of lead.

Figure 6.7 shows a summary for range of penetration of the three types of radiation.

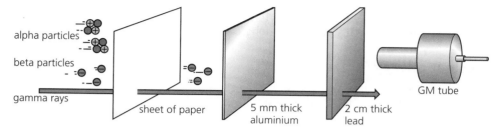

Figure 6.7 The penetrative range of the three types of radiation

Nuclear fission

Some heavy nuclei, such as uranium, can be forced to split into two lighter nuclei. The process is called **nuclear fission**. It occurs when a uranium nucleus is struck by a slow neutron (Figure 6.8). The heavy nucleus splits and the fragments move apart at very high speed, carrying with them a vast amount of energy. At the same time, two or three fast neutrons are also emitted. These are the fission neutrons and they go on to produce further fission and so create a chain reaction.

In a nuclear power station, steps are taken to ensure that, on average, just one of the fission neutrons goes on to produce further fission. This is **controlled nuclear fission**. The heat produced in the reaction is used to turn water into steam and drive a turbine to generate electricity. In a nuclear bomb there is no attempt to control the fission process (Figure 6.9).

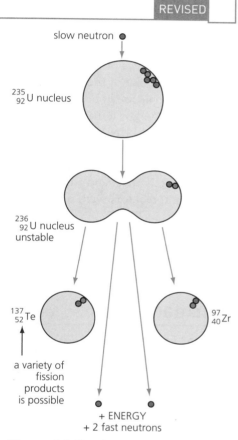

Figure 6.8 The fission of a uranium-235 nucleus

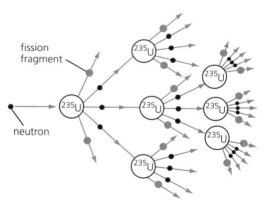

Figure 6.9 A chain reaction in uranium-235

A major disadvantage of all fission processes is that the fission fragments are almost always highly radioactive. This type of radioactive waste is *extremely dangerous* and expensive measures must be taken to store it until the level of activity is sufficiently small. In some cases this means the waste must be stored deep underground, in a vitrified (glass–like) state, for tens of thousands of years. One danger is that over time, the containers may leak and cause underground water pollution. Another danger is that earthquakes can rupture containers of radioactive waste buried underground, causing it to leak into the soil and water systems.

nuclear fission is the process by which a uranium (or plutonium) nucleus absorbs a slow neutron and then splits to produce two or more lighter nuclei and several neutrons, together with a vast quantity of energy

Issues relating to the use of nuclear energy to generate electricity

Nuclear energy is a controversial topic, which gives rise to strong political, social, environmental and ethical arguments on both sides of the debate.

Arguments in favour of nuclear energy

- It can produce vast amounts of energy/electricity.
- It produces very little carbon dioxide (CO_2) and hence does not contribute to global warming.
- It provides the 'base-load' for national electricity generation.
- It is a high-density source of energy.
- It provides employment opportunities for many people.
- Additionally, many nations are planning to build more nuclear reactors.

Arguments against nuclear energy

- The by-product of nuclear energy — nuclear waste and its disposal — has created one of the greatest problems of the twenty-first century.
- Many people are concerned about living close to nuclear power plants and the storage facilities used for radioactive waste. This gives rise to the acronym NIMBY (not in my back yard).
- This fear has been increased by
  - the disaster at Chernobyl in the Ukraine
  - the earthquake and tsunami in Fukushima, Japan in 2011, when several reactors were damaged leading to a meltdown and release of radiation.

Although nuclear fission does not release carbon dioxide, the mining, transport and purification of the uranium ore releases significant amounts of greenhouse gases into the atmosphere.

Nuclear fusion

REVISED

Nuclear fusion happens in stars like our Sun. At the centre of the Sun the temperature is about $15\,000\,000°C$. At this temperature, the nuclei of all atoms are stripped of their orbiting electrons and they are moving at a tremendous speed. Being positively charged, the nuclei would normally repel each other, but if they are moving fast enough, they can join (or fuse) to form a new nucleus (Figure 6.10). This causes the release of a vast amount of energy.

> **nuclear fusion** is the process in which light nuclei such as hydrogen combine together to produce a heavier nucleus such as helium and emit a vast quantity of energy

The equation representing the fusion process is:

$$^{2}_{1}H + \,^{3}_{1}H \;\rightarrow\; \,^{4}_{2}He + \,^{1}_{0}n + \text{energy}$$

Difficulties with fusion

The main problem is how to *contain* the reacting plasma at a *high enough temperature* and at *high particle densities* for a sufficiently *long time* for the reaction to take place.

A large advantage of fusion is that the isotopes of hydrogen — deuterium and tritium — are widely available as the constituents of sea water and so are nearly inexhaustible. Furthermore, fusion does not emit carbon dioxide or other greenhouse gases into the atmosphere since its major by-product is helium, an inert non–toxic gas.

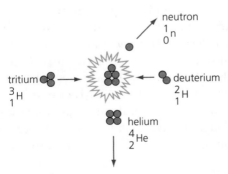

Figure 6.10 The fusion of deuterium and tritium

Fusion versus fission

Fusing nuclei together in a controlled way releases four million times more energy per kg than a chemical reaction such as the burning of coal, oil or gas; and fusing nuclei together in a controlled way releases four times as much as nuclear fission reactions per kg of fuel.

Fusion: solution to world's energy crisis?

There are many difficulties to overcome before nuclear fusion provides electricity on a commercial scale and it may be another 50 years before that happens. Nuclear fusion reactors will be expensive to build, and the system used to contain the plasma will be equally expensive because of the very high temperatures needed for the nuclei to fuse.

What is ITER?

REVISED

ITER ('the way' in Latin) is one of the most ambitious energy projects in the world today. ITER stands for 'International Thermonuclear Experimental Reactor'.

In southern France, 35 nations are collaborating to build the world's largest tokamak, a magnetic fusion device that has been designed to prove the feasibility of fusion as a large-scale and carbon-free source of energy based on the same principle that powers our Sun and stars.

ITER will be the first fusion device to:
● produce net energy
● maintain fusion for long periods of time

ITER has been designed mainly to:
● produce 500 MW of fusion power
● demonstrate the integrated operation of technologies for a fusion power plant

> **Typical mistake**
>
> Many students spell fission with one 's' and/or fusion with 'ss'. Make sure you can spell these terms correctly to be sure of scoring the mark.

> **ITER** is the 'International Thermonuclear Experimental Reactor', an experimental nuclear fusion device being built in southern France and involving the cooperation of scientists and engineers from 35 nations.

Exam practice

1 What mass of nitrogen-13 would remain if 80 g were allowed to decay for 30 minutes? Nitrogen-13 has a half-life of 10 minutes.
2 How long would it take for 20 g of cobalt-60 to decay to 5 g? The half-life of cobalt-60 is 5.26 years.
3 Strontium-93 takes 32 minutes to decay to 6.25% of its original mass. Calculate the value of its half-life.
4 When a radioactive material of half-life 24 hours arrives in a hospital its activity is 1000 Bq. Calculate its activity 24 hours before and 72 hours after its arrival.
5 (a) The table below shows the particles that make up a neutral carbon atom. Copy and complete the table showing the mass, charge, number and location of the particles. Some information has already been added for you. [7]

Particle	Mass	Charge	Number	Location
Electron		−1		
Neutron	1		6	in the nucleus
Proton			6	

(b) Radon is a naturally occurring radioactive gas.
(i) Explain what is meant by radioactive. [2]
(ii) Explain the danger of breathing radon gas into the lungs. [2]
(iii) Explain, in terms of the particles that make up the nucleus, the meaning of *isotope*. [2]

6 $^{12}_{6}C$ and $^{14}_{6}C$ are both isotopes of carbon.

(a) (i) Write down one similarity about the nucleus of each isotope. [1]
(ii) Write down one difference in the nucleus of these isotopes. [1]

(b) $^{14}_{6}C$ is radioactive. It decays to nitrogen by emitting a beta particle.

Copy and complete the equation below which describes the decay.

$$^{14}_{6}C \rightarrow \, ^{?}_{?}N + \, ^{?}_{?}\beta$$ [4]

(c) $^{14}_{6}C$ is present in all living materials and in all materials that have been alive. It decays with a half-life of 6000 years. [2]

(i) Explain the meaning of the term *a half-life of 6000 years*.
(ii) The activity of a sample of wood from a freshly cut tree is measured to be 80 disintegrations per second. Estimate the decrease in activity of the sample after 3 half-lives. [3]

7 A certain material has a half-life of 12 minutes. What proportion of that material would you expect still to be present an hour later? [3]

8 Four unknown nuclei are labelled W, X, Y and Z. Their full symbols are given below.

$$^{30}_{15}W \qquad\qquad ^{30}_{16}X \qquad\qquad ^{32}_{17}Y \qquad\qquad ^{33}_{16}Z$$

(a) Which, if any, of these nuclei are isotopes of the same element? [1]
(b) Explain your answer to part (a). [2]

9 (a) Describe the process of nuclear fusion. Your description should include:
● the particles involved
● what happens when nuclear fusion takes place
● where nuclear fusion occurs naturally. [6]

(b) A great deal of money is being invested on research into nuclear fusion.
(i) Suggest a reason why. [1]
(ii) Give two practical difficulties that must be overcome before fusion reactors become viable. [2]

(c) (i) What is ITER?
(ii) Explain the need for the ITER project. [3]

Answers online

ONLINE

7 Waves

Waves transfer **energy** from one point to another but they do not, in general, transfer matter. *All* waves are produced as a result of **vibrations** and can be classified as longitudinal or transverse.

Longitudinal waves

A **longitudinal wave** is one in which the particles vibrate parallel to the direction in which the wave is travelling. The only types of longitudinal waves relevant to your GCSE course are:
- sound waves
- ultrasound waves
- slinky spring waves
- P-type earthquake waves.

a **longitudinal wave** is one in which the particles vibrate parallel to the direction in which the wave is travelling

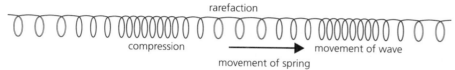

Figure 7.1 A longitudinal wave moving along a slinky spring

It is easy to demonstrate longitudinal waves by holding a slinky spring at one end and moving your hand backwards and forwards parallel to the axis of the stretched spring (Figure 7.1). **Compressions** are places where the coils (or particles) bunch together. **Rarefactions** are places where the coils (or particles) are furthest apart. All longitudinal waves are made up of compressions and rarefactions.

compressions are places where the coils (in a slinky) or particles in a longitudinal wave bunch together

rarefactions are places where the coils (in a slinky) or particles in a longitudinal wave are furthest apart

Transverse waves

A **transverse wave** is one in which the vibrations are at 90° to the direction in which the wave is travelling. Most waves in nature are transverse — some examples are:
- water waves
- slinky spring waves (Figure 7.2)
- waves on strings and ropes
- electromagnetic waves.

a **transverse wave** is one in which the particles vibrate perpendicular to the direction in which the wave is travelling

A transverse wave pulse can be created by shaking one end of a rope. The pulse moves along the rope, but the final position of the rope is exactly the same as it was at the beginning. None of the material of the rope has moved permanently. But the wave pulse has carried energy from one point to another.

Water waves are clearly transverse. A cork floating on the surface of water bobs up and down as the waves pass.

Exam tip

Slinky springs can be used to demonstrate both longitudinal *and* transverse waves — so it is best not to quote it if you are asked to give an example of a longitudinal (or a transverse) wave.

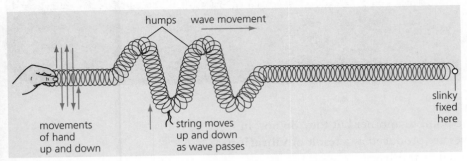

Figure 7.2 Transverse waves travelling through a slinky

Describing waves

There are a number of important definitions relating to waves that must be learned.

The **frequency** of a wave is the number of complete waves passing a fixed point in a second. Frequency is given the symbol f, and is measured in units called hertz (abbreviation Hz).

The **wavelength** of a wave is the distance between two consecutive crests or troughs (Figure 7.3). Wavelength is given the symbol λ, and is measured in metres. λ is the Greek letter 'l' and is pronounced 'lamda'.

The **amplitude** of a wave is the greatest displacement of the wave from its undisturbed (equilibrium) position (Figure 7.3). Amplitude is measured in metres.

> **frequency** is the number of complete waves passing a fixed point in a second
>
> **wavelength** is the distance between two consecutive crests or troughs
>
> **amplitude** is the greatest displacement of the wave from its undisturbed position

Figure 7.3 Displacement–distance graph to illustrate wavelength and amplitude

Wavelength and amplitude of longitudinal waves

For a **longitudinal** wave, the wavelength is the distance between the centre of one compression and the next. The amplitude of a longitudinal wave is the maximum distance a particle moves from the centre of this motion (Figure 7.4).

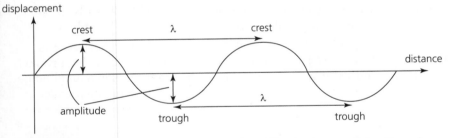

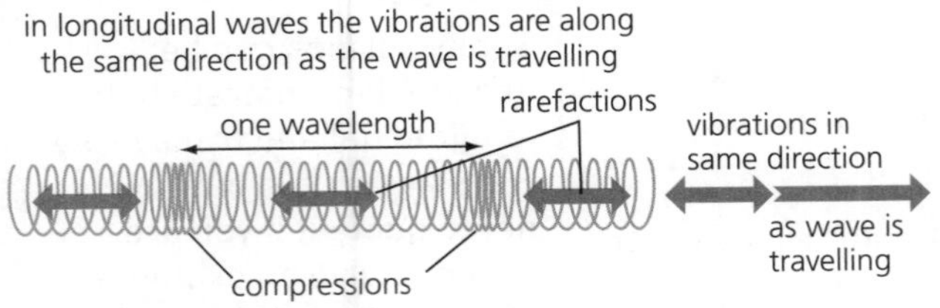

Figure 7.4 In longitudinal waves, the vibrations are along the same direction as the wave is travelling

Answers at **www.hoddereducation.co.uk/myrevisionnotesdownloads**

Now test yourself

1 Describe the difference between a transverse wave and a longitudinal wave.
2 Give two examples of transverse waves and two examples of longitudinal waves.
3 Define the terms *wavelength*, *frequency* and *amplitude* and state a unit in which each could be measured.
4 What evidence can you give that microwaves transmit energy?
5 How could you use a slinky spring to demonstrate:
   (a) a longitudinal wave
   (b) a transverse wave?
6 What happens to the compressions in a longitudinal wave on a slinky spring if the wavelength is increased?

Answers online

The wave equation

Imagine a wave with wavelength λ (metres) and frequency f (hertz). Then the speed of the wave, v, is given by:

wave speed = frequency × wavelength

$v = f \times \lambda$

> **Exam tip**
>
> This important equation must be learned for the GCSE examination.

Note that the units used in the wave equation must be consistent, as shown in Table 7.1.

Table 7.1

Frequency	Wavelength	Speed
always in Hz	cm	cm/s
	m	m/s
	km	km/s

Graphs and waves

We can represent waves on graphs like those shown in Figure 7.5.

Note carefully: The upper graph is displacement against **time**. The lower graph is displacement against **distance**.

The vertical axis in both cases is displacement — so we can find the amplitude from either graph.

The red line in the upper graph shows the time, T, between the crests passing a fixed point. This time is known as the period.

Students often wrongly think T is a distance (the wavelength).

The blue line in the lower graph shows the distance between consecutive crests. This is the wavelength, λ.

The period, T, in the upper graph is 4 seconds, and the wavelength λ is 10 metres.

How could we find the speed from these data?

The graphs tell us that the wave travels 10 metres in 4 seconds.

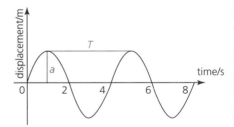

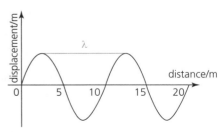

Figure 7.5 Graphs showing displacement against time and displacement against distance

$$\text{speed} = \frac{\text{distance}}{\text{time}} = \frac{10\,\text{m}}{4\,\text{s}} = 2.5\,\text{m/s}$$

$$\text{frequency}, f = \frac{1}{T} = 0.25\,\text{Hz}$$

We can use the frequency to confirm the speed using the wave equation:

$$v = f \times \lambda = 0.25\,\text{Hz} \times 10\,\text{m} = 2.5\,\text{m/s}$$

Now test yourself

7 Copy and complete the table below. Note carefully the unit in which you are to give your answers. The first one has been done for you as an example.

Wavelength	Frequency	Speed
5 m	100 Hz	500 m/s
12 m	50 Hz	_____ m/s
3 cm	60 kHz	_____ m/s
_____ m	4 Hz	20 cm/s
_____ m	5 kHz	2.5 km/s
16 mm	_____ Hz	80 cm/s
6×10^4 m	_____ Hz	3×10^8 m/s

8 The vertical distance between a crest and a trough is 24 cm and the horizontal distance between the first and the fifth wave crests is 40 cm. If 30 such waves pass a fixed point every minute, find the amplitude, frequency, wavelength and speed of the waves.

Answers online

Reflection

Figure 7.6 shows some plane waves approaching a straight metal barrier. The barrier is big enough to prevent waves going 'over the top'. The incident waves are **reflected** from the barrier.

> **reflection** is sending back a wave into the medium from which it came

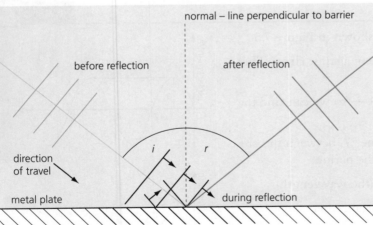

Figure 7.6 The reflection of waves off a plane surface. The angle of incidence = i, the angle of reflection = r

Note carefully that:
- the angle of incidence always equals the angle of reflection
- the wavelengths of incident and reflected waves are equal
- the frequency of the incident waves is the same as that of the reflected waves
- there is continuity of incident waves and reflected waves at the barrier.

Answers at **www.hoddereducation.co.uk/myrevisionnotesdownloads**

F The behaviour of water waves at a boundary is very similar to that of light at a mirror. However, water waves can be observed easily because they have a wavelength of many centimetres. Light waves have a wavelength typically around half a millionth of a metre, so their wave behaviour is a little more difficult to demonstrate.

Refraction

In Figure 7.7, water waves are travelling from deep water to shallow water.

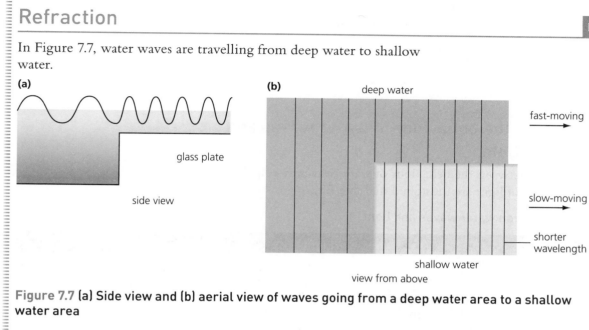

Figure 7.7 (a) Side view and **(b)** aerial view of waves going from a deep water area to a shallow water area

Waves travel more slowly in shallow water than they do in deep water. Since the same number of waves leave the deep water as enter the shallow water every second, the frequencies in the deep and shallow regions must be the same. This in turn means that the waves in shallow water must have a shorter wavelength than those in deep water.

When water waves enter the shallow region obliquely (at an angle), they not only slow down, but also change direction, as shown in Figure 7.8.

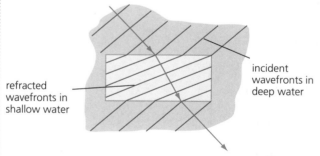

Figure 7.8 Waves changing direction as they pass from deep to shallow water

Note carefully that:
- the angle of incidence in deep water is always bigger than the angle of **refraction** in shallow water
- the wavelength and speed of waves in deep water are greater than those in shallow water
- the frequencies of waves in both deep and shallow water are the same
- there is continuity of incident and refracted waves at the boundary.

> **refraction** is the change in direction of a wave passing from one medium to another caused by its change in speed

A deep water wave of wavelength 12 cm and speed 36 cm/s enters a shallow region where the wavelength is 8 cm. Find the frequency and wave speed in the shallow water.

Answer

frequency in shallow water = frequency in deep water = $\dfrac{v}{\lambda}$

$$= \dfrac{36\,\text{cm/s}}{12\,\text{cm}} = 3\,\text{Hz}$$

speed in shallow water = $f \times \lambda = 3\,\text{Hz} \times 8\,\text{cm} = 24\,\text{cm/s}$

Analogy between the behaviour of water waves and the behaviour of light

The many similarities between the behaviour of water waves in a ripple tank and the behaviour of light are illustrated in Table 7.2.

Table 7.2 Analogy between water waves and light.

Water waves	Light
When they reflect: ● angle of incidence = angle of reflection ● reflected wavelength = incident wavelength ● reflected frequency = incident frequency ● reflected speed = incident speed	When it reflects: ● angle of incidence = angle of reflection ● reflected wavelength = incident wavelength ● reflected frequency = incident frequency ● reflected speed = incident speed
When they pass from deep into shallow water: ● they bend towards the normal ● refracted wavelength is less than incident wavelength ● refracted frequency = incident frequency ● refracted speed is less than incident speed	When it passes from air into glass and it refracts: ● it bends towards the normal ● refracted wavelength is less than incident wavelength ● refracted frequency = incident frequency ● refracted speed is less than incident speed

The close analogy between water waves and light waves enabled physicists to claim with some confidence that light waves, like water waves, are transverse.

Exam tip

Make sure you know the following:
● the wave equation and the units in which speed, wavelength and frequency are measured
● how to find the frequency of a wave from a graph of displacement against time
● how to find the wavelength of a wave from a graph of displacement against distance
● how to describe the analogy between reflection and refraction of water waves and the reflection and refraction of light.

Now test yourself

F

9 Figure 7.9 shows three wavefronts in a ripple tank approaching a solid barrier. The barrier acts as a reflector.

(a) Copy and complete the diagram to show what happens when the waves are reflected.

(b) In what way, if at all, do the frequency, wavelength and speed of water waves change when they are reflected?

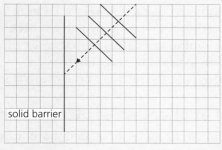

solid barrier

Figure 7.9

10 Figure 7.10 shows three wavefronts in deep water in a ripple tank. The water to the left of the vertical line is shallow water.

(a) Copy and complete the diagram to show what happens when the waves are refracted.

(b) In what way, if at all, do the frequency, wavelength and speed of these water waves change when they are refracted?

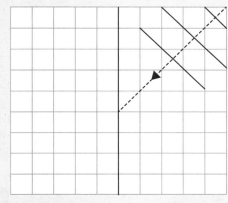

Figure 7.10

11 (a) The behaviour of water waves is analogous to that of light. What does this mean?

(b) State three properties of water waves those are analogous to those of light waves.

Answers online

Echoes

Like all waves, sound and ultrasound can be made to reflect. When this happens the angle of incidence is always equal to the angle of reflection (Figure 7.11).

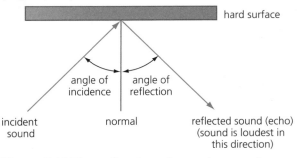

hard surface

angle of incidence

angle of reflection

incident sound

normal

reflected sound (echo) (sound is loudest in this direction)

Figure 7.11 The reflection of sound waves from a surface

Audible sound ranges in frequency from 20 Hz to 20 000 Hz. Sound above 20 000 Hz is called ultrasound and cannot be heard by humans. It can, however, be detected by bats, dogs, dolphins and many other animals.

Reflected sound (and ultrasound) is called an echo. Humans have found clever ways to use ultrasound echoes:

- scanning metal castings for faults or cracks (e.g. in rail tracks)
- scanning a woman's womb to check on the development of her baby
- scanning soft tissues to diagnose cancers
- fish location by seagoing trawlers
- mapping the surface of the ocean floor in oceanography.

An application of ultrasound in medicine

In an ultrasound scan of an unborn baby, a probe is moved across the mother's abdomen. The probe sends out ultrasound waves and also detects the reflections. The other end of the probe is connected to a computer.

By examining the reflected waves from the womb, the computer builds up a picture of the foetus (unborn baby). Unlike X-rays, ultrasound is quite safe for this purpose.

Ultrasound can also be used to measure the diameter of the head of the baby as it develops in the womb (Figure 7.12). When the ultrasound reaches the baby's head at A, some ultrasound is reflected back to the detector and produces pulse A on the cathode ray oscilloscope (CRO). Some ultrasound passes through the head to point B, and is then reflected back to the detector. This reflection produces pulse B on the CRO.

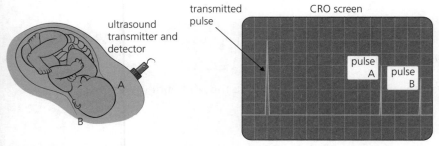

Figure 7.12 The diameter of a baby's head can be measured using ultrasound

In the diagram of the CRO screen, each horizontal division corresponds to a time of 40 microseconds ($40\,\mu s = 40 \times 10^{-6}\,s$).

The time interval between the arrival of pulse A and the arrival of pulse B at the detector corresponds to three divisions on the CRO. Since each division is $40\,\mu s$ this represents a total time of $120\,\mu s$. This additional $120\,\mu s$ is the time taken for ultrasound to travel from A to B and back to A. The time for ultrasound to travel from B to A is therefore half of that or $60\,\mu s$.

Now physicists know that ultrasound travels at a speed of $1500\,m/s$ in a baby's head. So the width of the head can be found as follows:

width of head = speed × time = $1500\,m/s \times 60 \times 10^{-6}\,s = 0.09\,m = 9\,cm$

Scanning metal castings

Railway tracks do not last forever. They wear out. So it is important that we find out early if they are developing cracks or flaws below the surface.

We can do this with ultrasound scanners attached to specially fitted rail carriages (Figure 7.13). Ultrasound passes through the track. If there is a crack or other flaw it can be imaged (using the same science that allows us to obtain a picture of a baby in the womb).

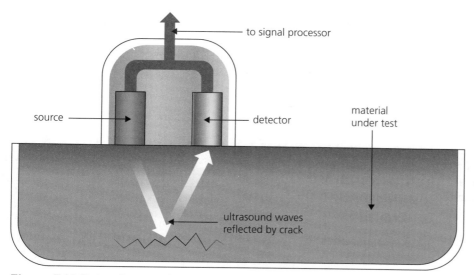

Figure 7.13 Detecting cracks in metals

Sonar

Sonar stands for **SO**und **N**avigation **A**nd **R**anging and was originally developed to detect submarines in the early twentieth century. However, the following example illustrates its use by fishermen to detect shoals of fish and to measure how far they are below the surface.

Example

A fishing trawler produces an ultrasound pulse and 0.4 s after it is transmitted an echo from a shoal of fish is detected. Assuming the speed of ultrasound in seawater is 1500 m/s:
(a) calculate the total distance travelled by the ultrasound
(b) calculate the distance from the trawler to the shoal of fish
(c) explain why a second echo is detected shortly after the first.

Answer

(a) total distance = speed × time = 1500 m/s × 0.4 s = 600 m
(b) distance from the trawler to the fish = ½ × total distance
    = ½ × 600 m = 300 m
(c) The first echo is from the fish. The second echo is from the sea bed, which is further away from the trawler than the fish. So the ultrasound from the sea bed takes longer to reach the trawler than the echo from the fish.

Radar

Radar stands for **RA**dio **D**etection **A**nd **R**anging and was originally developed during World War II to detect enemy aircraft and find their distance from the radar station.

Radar waves are in the microwave section of the electromagnetic spectrum. Think of a radar beam as a powerful beam of microwaves. They have wavelengths ranging from a few mm to just over a metre. Because radar waves are incredibly fast at 300 000 000 m/s, they are used extensively to track very fast objects that may be a large distance away. So, for example, they are used by air traffic controllers to track passenger airliners, by the military to track missiles and by the coastguard to detect ships.

Radar cannot be used under water. The water absorbs the radar (microwaves) within a metre or so. However, the physics of the application is very similar to that of sonar. If you are asked to solve a mathematical problem involving radar it is likely that numbers will be given in index form as shown in the example.

Example

Figure 7.14 shows a ground-based radar station A. It transmits a radar beam, which reflects off aircraft B. The radar echo is received at A 2.2×10^{-4} s after the original radar transmission. If the speed of radar waves is 3×10^{8} m/s, calculate the distance AB in km.

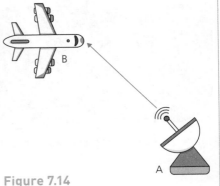

Answer

The time taken for radar to travel from A to B and back to A is 2.2×10^{-4} s

So, the time taken for radar to travel from A to B is $\frac{1}{2} \times 2.2 \times 10^{-4}$ s

$$= 1.1 \times 10^{-4}\,\text{s}$$

Distance AB = speed × time = 3×10^{8} m/s × 1.1×10^{-4} s

$$= 3.3 \times 10^{4}\,\text{m} = 33\,\text{km}$$

Figure 7.14

Electromagnetic waves

Electromagnetic waves are members of a family with common properties called the electromagnetic spectrum. They:
- can travel in a vacuum (unique property of electromagnetic waves)
- travel at exactly the same speed in a vacuum
- are transverse waves.

Electromagnetic waves also show properties common to all types of wave. They:
- carry energy
- can be reflected
- can be refracted.

There are seven members of the electromagnetic family. The properties of electromagnetic waves depend very much on their wavelength. In Table 7.3 they are arranged in order of increasing wavelength (or decreasing frequency). CCEA students need to be able to list these waves in order of increasing (and decreasing) wavelength. But you do not need to remember the wavelengths! Table 7.4 lists some dangers of electromagnetic waves.

Table 7.3 The electromagnetic spectrum

Electromagnetic wave	Typical wavelength
Gamma (γ) rays	0.01 nm*
X-rays	0.1 nm
Ultraviolet light	10 nm
Visible light	500 nm
Infrared light	0.01 mm
Microwaves	3 cm
Radio waves	1000 m

*1 nanometre (nm) = 1×10^{-9} m

Table 7.4 Dangers of electromagnetic waves

Electromagnetic wave	Dangers
Gamma (γ) rays	Damage cells and disrupt DNA, which may lead to cancer
X-rays	Damage cells and disrupt DNA, which may lead to cancer
Ultraviolet light	Certain wavelengths can damage skin cells, disrupting DNA and leading to skin cancer
Visible light	Intense visible light can damage the eyes (e.g. snow blindness)
Infrared light	Felt as heat and can cause burns
Microwaves	Cause internal heating of body tissues which, some say, can lead to eye cataracts
Radio waves	Large doses of radio waves are believed to cause cancer, leukaemia and other disorders and some people claim the very low-frequency radio waves from overhead power cables near their homes has affected their health

Now test yourself

The following question shows that while you do not need to remember the wavelengths of the members of the electromagnetic spectrum, questions on their wavelengths can be asked.

12 Below are three members of the electromagnetic spectrum, not arranged in any particular order.

   gamma (γ) rays   radio waves   ultraviolet light

Copy and complete the table below, writing the name of the missing electromagnetic wave opposite its typical wavelength.
(Hint: first identify the missing members of the spectrum and write them in order of increasing wavelength.)

Wave				
Typical wavelength/m	1×10^{-10}	6×10^{-7}	1×10^{-5}	1×10^{-3}

13 The emitter in Figure 7.15 sends out a pulse of sound. An echo from the object is detected after 2.5 ms. If sound travels at 340 m/s in the air, calculate the distance marked *d*.

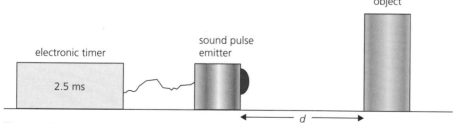

Figure 7.15

Answers online

Exam practice

1 What physical property of a water wave never changes as a result of:
   (a) reflection
   (b) refraction? [2]
2 The graphs in Figure 7.16 relate to a water wave.

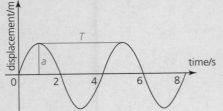 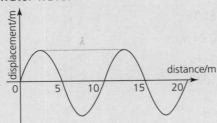

Figure 7.16

Find the frequency, wavelength and speed of the wave. [6]
3 (a) What is ultrasound? [1]
   (b) In what way is ultrasound different from the sound of human speech? [1]
   (c) State two differences between ultrasound waves and electromagnetic waves. [2]

The following question is provided to give you practice at the kind of question that could appear in Part B of Unit 3 (Part B of Unit 7 in Double Award Science).
4 When the frequency of sound is changed, the wavelength also changes. The table below shows the results of an experiment to measure the wavelength of sound at different frequencies. The unit for $1/\lambda$ is 1/m.

Wavelength, λ/m	0.7	1.0	1.5	2.5	4.0
Frequency, f/Hz	460	320	210	130	80
$1/\lambda$				0.40	0.25

   (a) Complete the table by entering the missing numbers in the third row. Two entries have already been done for you. [3]
   (b) Plot (on graph paper) the graph of f/Hz on the vertical axis against $1/\lambda$ on the horizontal axis and draw the straight line of best fit. [4]
   (c) Find the gradient of your line of best fit and state its unit. [3]
   (d) What is the physical significance of the gradient of the line of best fit? [1]
   (e) Use your graph to find the wavelength of sound of frequency 250 Hz. [2]

Answers online

ONLINE

8 Light

Reflection of light

REVISED

Figure 8.1 shows the reflection of light from a straight plane mirror.

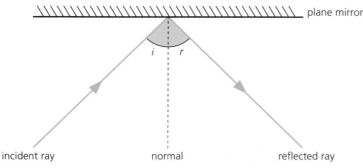

Figure 8.1 **Reflection by a plane mirror**

Experiments show that the **angle of incidence** is always *equal to* the **angle of reflection**. This is known as the **law of reflection**.

You should be able to describe the following experiment to demonstrate this law.

> **angle of incidence** (*i*) is the angle between the normal and the incident ray
>
> **angle of reflection** (*r*) is the angle between the normal and the reflected ray

Practical

The law of reflection

1 With a sharp pencil and a ruler, draw a straight line AOB on a sheet of white paper.
2 Use a protractor to draw a normal, N, at point O.
3 With the protractor, draw straight lines at various angles to the normal ranging from 15° to 75°.
4 Place a plane mirror on the paper so that its back rests on the line AOB (Figure 8.2).
5 Using a ray box, shine a ray of light along the line marked 15°.
6 Mark two crosses on the reflected ray on the paper.
7 Remove the mirror, and using a ruler join the crosses on the paper with a pencil and extend the line backwards to point O — this line shows the reflected ray.
8 Measure the angle of reflection with a protractor.
9 Record in a table the angles of incidence and reflection.
10 Repeat the experiment for different angles of incidence up to 75°.

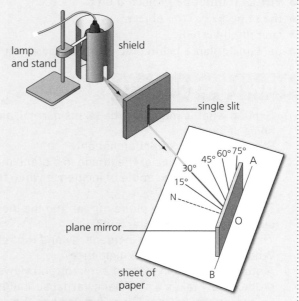

Figure 8.2 **Experiment to demonstrate the law of reflection**

Practical

Locating the image position in a plane mirror

1 Support a plane mirror vertically on a sheet of white paper, and with a pencil draw a straight line at the back to mark the position of the reflecting surface.
2 Use a ray box to direct two rays of light from point O towards points A and B on the mirror.
3 Mark the position of point O with a cross using a pencil (Figure 8.3).
4 Mark two crosses on each of the real reflected rays.
5 Remove both the ray box and the mirror.
6 Using a ruler, join the crosses with a pencil line so as to obtain the paths of the real rays from A and B.
7 Extend these lines behind the mirror (these are called virtual rays) — they meet at I, the point where the image was formed.
8 Measure the distance from the image I to the mirror line (IN) and the distance from the object O to the mirror line (ON) — they should be the same.
9 Repeat the experiment for different positions of the object O.
10 In each case, the object O and its image I should be the same perpendicular distance from the mirror.

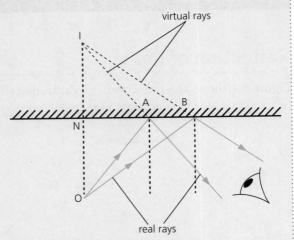

Figure 8.3 The image in a plane mirror is the same distance behind as the object is in front

Remember: the image in a plane mirror is:
- virtual (cannot be projected on to a screen)
- the same size as the object
- laterally inverted
- the same distance behind the mirror as the object is in front of the mirror.

Now test yourself

1 Explain what is meant by the terms *normal*, *angle of reflection* and *angle of incidence*.
2 State the law of reflection of light.
3 State the properties of the image in a plane mirror.
4 State the size of the angle of incidence when the incident ray strikes a plane mirror at 90°.
5 The angle between a plane mirror and the incident ray is 40°. What size is the angle of reflection?
6 The angle between the incident ray and the reflected ray is 130°. What size is the angle of incidence?
7 A student stands in front of a mirror and views his image. The student now takes a step backwards, so that he is 20 cm further away from the mirror. By how much has the distance between the student and his image in the mirror increased?

Answers online

Exam tip

Be very precise when asked about the image size and position in a plane mirror.

Don't just write: 'same size'. Remember to add... 'as the object'.

Don't just write: 'same distance behind the mirror'. Remember to add... 'as the object is in front of the mirror'.

Refraction of light

Refraction is the change in direction of a beam of light as it travels from one material to another due to a change in speed in the different materials (Figure 8.4). Table 8.1 shows the speed of light in various media. It is not necessary to remember the numbers in this table, but you must know that light travels faster in air than in water, and faster in water than in glass. In fact, the *greater* the change in the speed, the *greater* is the bending.

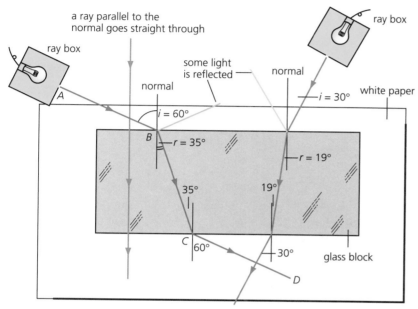

Figure 8.4 Refraction of light rays by a glass block

Table 8.1

Material	Speed of light in m/s
Air (or vacuum)	300 000 000
Water	225 000 000
Glass	200 000 000

Remember:
- The angle between the normal and the incident ray is the angle of incidence.
- The angle between the normal and the refracted ray is the **angle of refraction**.

Experiments show that:
- when light speeds up, it bends away from the normal
- when light slows down, it bends towards the normal.

Remember that this is also what happens to waves travelling from deep water into shallow water.

> **angle of refraction** is the angle between the normal and the refracted ray

Figure 8.5 shows what happens when light travels from air through glass, then through water, and finally back into the air. Note the changes of direction in each case.

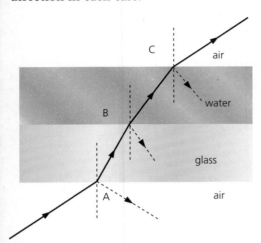

Figure 8.5 Refraction

Prescribed practical P6

Measuring angles of incidence and refraction

In this required practical activity you will measure angles of incidence and refraction as a ray of light passes from air into glass.

Apparatus
- rectangular glass (or Perspex) block
- ray box
- low-voltage power supply (PSU)
- leads
- protractor
- A4 plain white paper
- pencil
- ruler

Method
1. Prepare a table for your results like that shown on page 75.
2. Place the rectangular glass block in the centre of the sheet of white paper on a drawing board and draw round its outline with a sharp pencil.
3. Switch on the PSU and direct a ray of light to enter the block near the middle of the longest side of the block, so that the angle of incidence is about 10° (Figure 8.6).
4. Mark two pencil dots on the paths of both the incident ray and the emergent ray and remove the glass block.
5. Join the dots on the incident ray up to the point of incidence and join the dots on the emergent ray back to the point of emergence.
6. Draw a straight line between the point of incidence and the point of emergence.
7. Draw the normal at the point of incidence.
8. Measure the angle of incidence, i, and the angle of refraction, r, and record the data in your table.
9. Repeat steps 2 to 8 for angles of incidence ranging from about 20° up to about 80°.

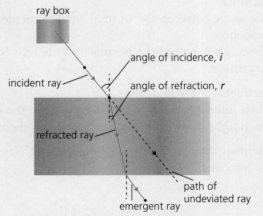

Figure 8.6 Light rays entering a glass block

Answers at **www.hoddereducation.co.uk/myrevisionnotesdownloads**

Results

Angle of incidence, $i/°$	10	20	29	41	49	58	67	79
Angle of refraction, $r/°$	7	13	19	26	30	34	38	41

The above results are typical of those obtained in this experiment.

Treatment of the results

Plot the graph of angle of incidence (y-axis) against angle of refraction (x-axis).

Conclusion

The graph of angle of incidence against angle of refraction (Figure 8.7) is a curve through the origin of increasing gradient (the graph gets steeper as the angle of incidence increases). This tells us that i is *not* directly proportional to r (because the graph is *not* a straight line), but that i and r have a positive correlation (as i increases, r increases).

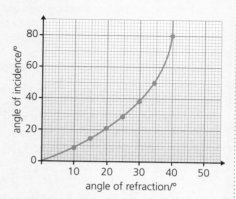

Figure 8.7 A graph showing angle of incidence against angle of refraction

---

Exam tip

Always remember to put an arrow on real rays of light. Both the normal and the virtual rays are always dotted and never have an arrow.

Dispersion of white light

All colours (frequencies) of light travel at the same speed in air. But different colours of light travel at different speeds in glass. This means that different colours bend by different amounts when they pass from air into glass (Figure 8.8). When light is passed through a triangular glass block, a prism, the effect is called **dispersion** and it results in a spectrum showing all the colours of the rainbow. Red light is bent (refracted) the least because it travels fastest in glass. Violet light bends the most because it is slowest in glass.

dispersion is the splitting of white light into its component colours

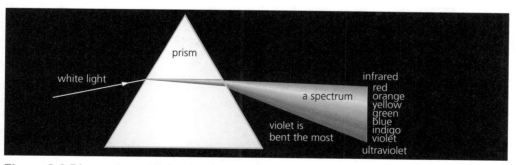

Figure 8.8 Dispersion of light through a prism

Now test yourself

8 Describe what is meant by refraction of light.
9 Explain what has happened to the speed of a ray of light if it refracts:
   (a) towards the normal
   (b) away from the normal.
10 (a) State what is meant by dispersion of light.
   (b) Describe the conditions necessary for dispersion to occur.
11 (a) Figure 8.9 shows a ray of light passing from the air into the cornea of the eye.

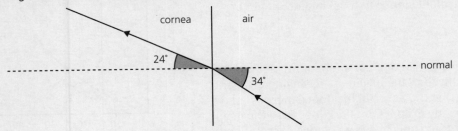

Figure 8.9

   (i) State the angle of incidence in air and the angle of refraction in the cornea.
   (ii) In which of the two media does light travel the faster?
   (b) What is the evidence for believing that red light travels faster in glass than blue light?
12 A sound wave made underwater travels towards the surface. Sound travels *faster* in water than it does in air. Copy and complete Figure 8.10 to illustrate the refraction that occurs when the sound travels into the air. Mark the normal on your diagram.

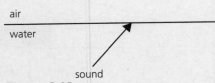

Figure 8.10

13 Different shapes of glass prism are often used to change the direction of light rays.

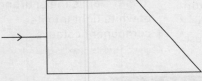

Figure 8.11

Copy Figure 8.11 and continue the path of the ray shown until it emerges into the air.

Answers online

⊕Critical angle

Figure 8.12 shows what happens when light travels through glass and emerges into the air. When the angle of incidence in glass is small enough, most of the light refracts into the air, but a little light is **internally** reflected.

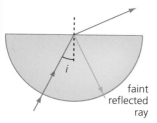
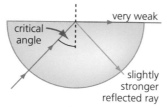
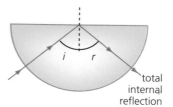

angle of incidence is **less** than the critical angle
Most of the light *passes through* into the air, but a little bit is *internally reflected*

angle of incidence is **equal** to the critical angle
The emerging ray comes out *along the surface*. There is quite a bit of *internal reflection*

angle of incidence is **greater** than the critical angle
No light comes out
It is *all* internally reflected, i.e. *total internal reflection*

Figure 8.12 Total internal reflection of light

As the angle of incidence increases, the refracted ray bends closer and closer to the glass and becomes weaker and weaker. At the same time, the light being internally reflected into the glass becomes stronger. Eventually, at a certain angle of incidence called the **critical angle**, the light is refracted at an angle of refraction of 90°. At this point the refracted ray is very weak and the internally reflected ray is quite strong.

The critical angle for glass is about 42°. At angles above the critical angle, there is no refraction at all. *All* the light is reflected back into the glass — this is called **total internal reflection**.

> **critical angle** is the angle of incidence in the medium resulting in an angle of refraction of 90° in air
>
> **total internal reflection** is what occurs when light is travelling in an optically dense material (like glass) towards a boundary with an optically rare material (like air), and the angle of incidence at the boundary is greater than the critical angle

> **Exam tip**
>
> Remember that:
> - the critical angle is the angle of incidence in a material for which the angle of refraction in air is 90°
> - at angles of incidence less than the critical angle, both reflection and refraction occur
> - at angles of incidence greater than the critical angle, no refraction occurs and the light is totally internally reflected.
> - for total internal reflection to occur, light must be travelling from an optically dense material like glass towards a material which is less optically dense (like air).

Practical

Finding the critical angle

Apparatus
- semicircular glass (or Perspex) block
- ray box
- low-voltage power supply (PSU)
- leads
- protractor
- A4 plain white paper
- pencil
- ruler

→

Method

1 Place the semicircular glass block in the centre of the sheet of white paper and draw round its outline with a sharp pencil.
2 Remove the block, mark the centre X of the straight diameter and then replace the block.
3 Switch on the PSU and direct a ray of light towards X, through the block, as in Figure 8.12. Ensure that the angle of incidence, i, is small so that light can be seen leaving the block at X.
4 Continue to direct the ray towards X, but slowly move the ray box so that the angle of incidence at X increases. Observe the internally reflected ray becomes stronger.
5 Continue to move the ray box to increase the angle of incidence at X, until the refracted ray at X just emerges along the diameter. Observe that if the angle of incidence at X is now increased even slightly, the light is totally internally reflected.
6 With a pencil draw two dots on the incident ray as far apart as possible and then remove the glass block.
7 Join the dots on the incident ray and extend the line beyond the point of incidence until it reaches X.
8 Draw the normal at X.
9 Measure the critical angle at point X and record the value in a table.
10 For reliability, repeat steps 1 to 9 about 4 more times and average the values found for the critical angle.

Total internal reflection in parallel-sided glass blocks

When the angle of incidence in glass is equal to the critical angle we just get refraction.

At angles of incidence above the critical angle we get total internal reflection.

> **Exam tip**
>
> The critical angle is the angle of incidence in glass for which the angle of refraction in air is 90°.

Optical fibres

REVISED

Optical fibres are lengths of solid glass core with an outer plastic sheath. Provided the fibre is not bent too tightly, light will strike the core–cladding boundary at an angle greater than the critical angle and be totally internally reflected at the surface of the glass core. However, every optical fibre has some imperfections at its reflecting surface and this means that the signal must be boosted every kilometre or so in communications links. Optical fibres are used to transmit both telephone and video signals over long distances.

The big advantage of optical fibres is that they can carry much, much more information than a copper cable of the same diameter.

If the optical fibre is too tightly bent, the angle of incidence at the core–cladding boundary may become less than the critical angle and light will be lost by refraction into the cladding.

Endoscopes are used by surgeons to look inside a patient's body without needing to cut a large hole. They consist of bundles of optical fibres that allow light to travel into the body and then allow image information to pass out of the body. The surgeon can therefore see on a monitor what is happening inside the body, as it happens. The endoscope kit also carries tools for cutting, snaring, water irrigation and retrieval of tissue. It is the use of optical fibres that makes keyhole surgery (laparoscopy) possible. Other examples of the use of total internal reflection include prism binoculars and the prism periscope.

Answers at **www.hoddereducation.co.uk/myrevisionnotesdownloads**

14 Define what is meant by the term *critical angle*.
15 State the two conditions required for total internal reflection to occur.
16 Describe, in detail, an experiment to measure the critical angle in a semicircular glass block.
17 Describe the structure of an optical fibre and state the applications of fibre optics to telecommunications and medical endoscopy.
18 A ray of light strikes an equilateral glass prism normally as shown in Figure 8.13. The critical angle for the glass is 42°.

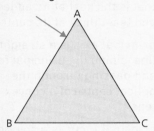

Figure 8.13

(a) State the angle of incidence in air at surface AB.
(b) Calculate the size of the angle at A and hence the size of the angle of incidence at the surface AC.
(c) State what happens to the light at surface AC.
(d) Calculate the angle of refraction in air at surface BC.
(e) Copy the diagram and complete it to show the passage of the light through the glass and back into the air.

19 Copy and complete the table by placing ticks (✔) in the correct boxes. There is no need to copy the diagrams.

	Is the statement true or false?	True	False
	No refraction is taking place at the curved surface.		
	The angle of incidence in air at the curved surface is 0°.		
	The angle marked *i* is less than the critical angle.		
	No refraction is taking place at point X.		
	Refraction in the air will occur when the angle marked *c* is increased.		
	The light is travelling faster in the semicircular block than it is in the air.		

20 Does light travelling in the core of an optical fibre travel faster, slower or at the same speed as light travelling in the cladding? Give a reason for your answer.

Answers online

Lenses

Lenses are specially shaped pieces of glass or plastic. There are two main types of lens:

- **converging** (or convex)
- **diverging** (or concave).

These are shown in Figure 8.14.

converging (convex) lens

converging lens is thickest at the centre

diverging (concave) lens

diverging lens is thickest at the edges

Figure 8.14 The shapes of a converging lens and a diverging lens

There is one feature of a converging lens that needs to be defined:
- Rays of light parallel to the **principal axis** of a converging lens all converge at the **principal focus** on the opposite side of the lens.

The distance between the principal focus and the optical centre of *any lens* is called the **focal length**.

For a convex lens, light refracts at each surface as it enters and leaves the lens, first bending towards the normal and then away from the normal.

There is one feature of a diverging lens that needs to be defined:
- Rays of light parallel to the principal axis of a diverging lens all appear to diverge from the principal focus after refraction in the lens.

Note that light passing through the optical centre of a convex or concave lens is not bent. It passes straight through without refraction.

> a **converging** lens is a lens that is thickest at its centre and least thick at its edges
>
> a **diverging** lens is a lens that is thickest at its edges and least thick at its centre
>
> **principal axis** is a straight line joining the principal foci and passing through the optical centre of a convex lens
>
> **principal focus** is a point on the principal axis of a convex lens through which rays of light parallel to the principal axis pass after refraction in the lens
>
> **focal length** is the distance between the optical centre of a lens and the principal focus

Practical

Measuring the focal length, *f*, of a converging lens using a distant object

Apparatus
- convex lens
- lens holder
- ruler
- sellotape
- white screen in a holder
- distant object (such as a tree, which can be seen through the windows in the laboratory and is at least 20 m away)

Method
1. Sellotape the ruler to the bench.
2. Place the white screen in its holder at the zero mark.
3. Place the lens in its holder as close as possible to the screen.
4. Slowly move the lens away from the screen until the inverted image of the distant object is as sharp as possible.
5. Using the metre ruler, measure the distance from the centre of the lens to the white screen. This distance is the focal length of the lens.
6. Record the measured focal length in a prepared table.
7. For reliability, repeat steps 2 to 6 for four different distant objects and determine the average value of *f*.

Results

Focal length, *f*/mm	245	240	250	247	243
Average focal length, *f*/mm	245				

Answers at **www.hoddereducation.co.uk/myrevisionnotesdownloads**

Image in a concave (diverging) lens

Note that regardless of the position of the object, the image in a concave (diverging) lens is *always*:

- erect
- virtual
- smaller than the object
- placed between the object and the lens.

This is shown in the ray diagram in Figure 8.15.

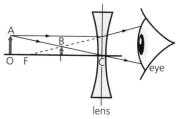

Figure 8.15 The image in a diverging lens

Image in a convex (converging) lens

The position and properties of the image in a convex (converging) lens depend on the position of the object. We can find those positions and image properties by drawing a ray diagram.

Rules for drawing ray diagrams

To draw a ray diagram for a convex lens you must draw at least two of the following rays:

- a ray parallel to the principal axis, refracted through the principal focus on the other side of the lens
- a ray through the optical centre of the lens that does not change its direction (does not refract)
- a ray through the principal focus on one side of the lens, which emerges so that it is parallel to the principal axis on the other side of the lens.

First steps when drawing a ray diagram for a convex lens:

1 using a ruler, draw a horizontal line to represent the principal axis and a vertical line for the lens.
2 Mark the position of the principal focus with a letter F, the same distance from the optical centre on each side of the lens.
3 Using a ruler, draw a vertical line touching the principal axis at the correct distance from the lens to represent the object.
4 Using a ruler, draw at least two of the three construction rays, starting from the top of the object.
5 Draw arrows on all rays to show the direction in which the light is travelling.

The point where the construction rays meet is at the top of the image. The bottom of the image lies vertically below on the principal axis.

To illustrate the process, consider the following example.

Example

An object 5 cm tall is placed 6 cm away from a converging lens of focal length 4 cm. Find the position and height of the image.

Answer

In Figure 8.16, circled numbers have been added to show the order in which the lines or rays have been drawn. These numbers are drawn for illustration only and are normally omitted from such ray diagrams.

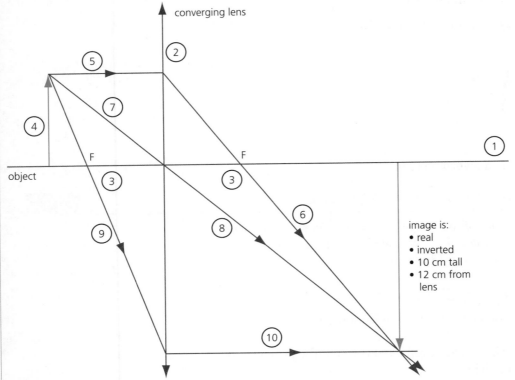

Figure 8.16 To work out the position and height of an image

① horizontal line representing principal axis (PA)
② vertical line representing lens
③ two principal foci marked, each 4 cm from lens
④ object marked 6 cm from lens
⑤ Ray from top of object parallel to PA...
⑥ ...refracts through F.
⑦ Ray from top of object through the optical centre...
⑧ ...is not refracted.
⑨ Ray from top of object through F...
⑩ ...refracts parallel to PA.

Finally the image is drawn from the point where the refracted rays meet to the PA.

The image is drawn as a continuous line to show that it is real (can be projected on to a screen).

The downward arrow on the image shows it is inverted.

Now use the ruler to measure the height of the image and its distance from the centre of the lens.

Ray diagrams

REVISED

The ray diagrams in Table 8.2 show where the image is formed for different positions of the object. You should study carefully the diagrams, and Table 8.3, which gives a summary of the information.

Answers at **www.hoddereducation.co.uk/myrevisionnotesdownloads**

Table 8.2 Drawing ray diagrams

Position of object	Ray diagram	Properties of image
(H) Between the principal focus, F, and the lens		● On same side of lens as object but further from lens ● Virtual ● Erect ● Larger than object
At principal focus, F		● Image is at infinity
Between F and twice the focal length		● On opposite side of lens to object but further away than twice the focal length ● Real ● Inverted ● Larger than object
At 2F		● On opposite side of lens to object and exactly same distance away as object ● Real ● Inverted ● Same size as object
Just beyond twice the focal length of the lens		● On opposite side of lens to object and between one and two focal lengths from lens ● Real ● Inverted ● Smaller than object

Table 8.3 Summary of ray diagrams

Position of object	Location of image	Properties of image			Application
		Nature	Erect or inverted	Larger or smaller than object	
(H) Between lens and F	On same side of lens as object but further away from lens	Virtual	Erect	Larger	Magnifying glass
At F	At infinity	Real	Inverted	Larger	Searchlight
Between F and 2F	Beyond 2F	Real	Inverted	Larger	Cinema projector
At 2F	At 2F	Real	Inverted	Same size	Telescope — erecting lens
Just beyond 2F	Between F and 2F	Real	Inverted	Smaller	Camera
Very far away from lens	At F	Real	Inverted	Smaller	Camera

Now test yourself

21 Draw a ray diagram to show how a diverging lens produces a virtual, diminished image.
22 State the rules for drawing ray diagrams for a convex lens.
23 Draw a ray diagram to illustrate how a converging lens can produce:
   (a) a real, magnified image
   (b) a diminished image
   (c) a virtual image.

Answers online

Exam tip

You should pay particular attention to the ray diagrams that illustrate the principles of:
● the magnifying glass (to give an erect, virtual image)
● the projector (to give a magnified, real image)
● the camera (to give a diminished real image).

They are specifically required by the subject specification.

The human eye

Figure 8.17 shows the human eye. While most refraction occurs in the cornea, the purpose of the lens is to produce a sharp image on the light-sensitive tissue called the retina.

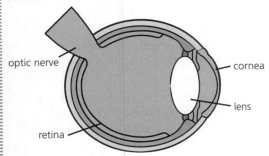

Figure 8.17 The eye

The normal eye

The farthest point that can be seen clearly by the eye is called the far point. For a normal eye this is at infinity. Light from the far point reaches the eye as parallel rays. The rays are refracted by the cornea and the lens so that they meet on the retina where they form a sharp image.

The nearest point that can be seen clearly by the unaided eye without causing the muscles to strain is called the near point. For a normal eye this is at 25 cm.

Short sight (myopia)

A person who suffers from **short sight (myopia)** is unable to see distant objects sharply. They cannot make the lens thin enough to view distant objects. This causes the light from distant objects to converge just in front of the retina (Figure 8.18a). The image seen by the person is blurred. To correct this defect a diverging lens is used (Figure 8.18b).

myopia (short sight) is a defect of the eye in which distant objects appear blurred because the images are focused in front of the retina rather than on it

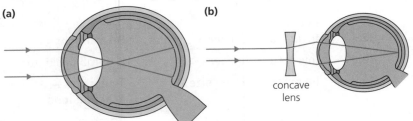

Figure 8.18 The use of a diverging (concave) lens to correct myopia

ⒻLong sight (hypermetropia)

A person suffering from **long sight (hypermetropia)** sees distant objects clearly but does not see near objects clearly. This happens because the patient's near point is much further than 25 cm (Figure 8.19a). To correct for this defect a converging lens is used (Figure 8.19b).

> **hypermetropia (long sight)** is a defect of the eye in which near objects appear blurred because the images are focused behind the retina rather than on it

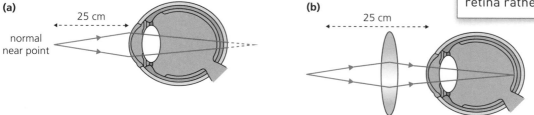

Figure 8.19 The use of a converging (convex) lens to correct hypermetropia

Exam practice

1 State four properties of the image in a plane mirror. [4]

2 Two mirrors, M_1 and M_2, are placed at right angles to one another. Figure 8.20 shows a ray of light incident on mirror M_1 at an angle of 27° to its surface.

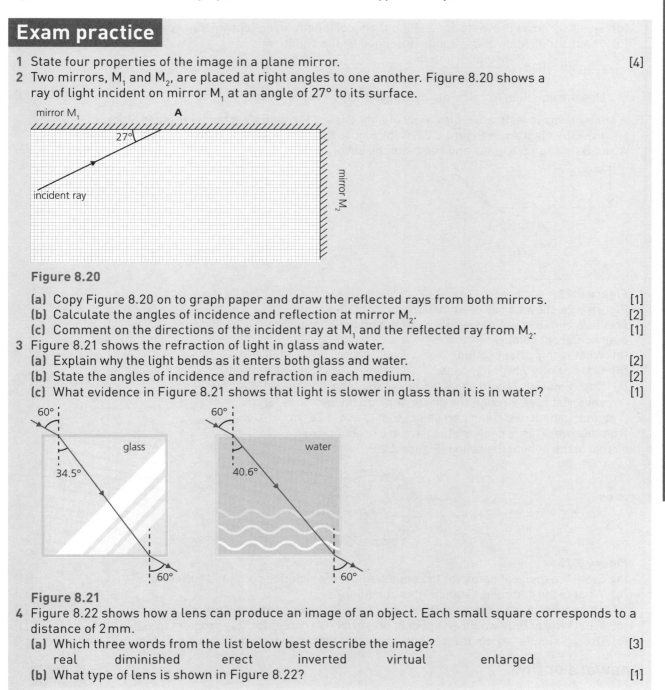

Figure 8.20

   (a) Copy Figure 8.20 on to graph paper and draw the reflected rays from both mirrors. [1]
   (b) Calculate the angles of incidence and reflection at mirror M_2. [2]
   (c) Comment on the directions of the incident ray at M_1 and the reflected ray from M_2. [1]

3 Figure 8.21 shows the refraction of light in glass and water.
   (a) Explain why the light bends as it enters both glass and water. [2]
   (b) State the angles of incidence and refraction in each medium. [2]
   (c) What evidence in Figure 8.21 shows that light is slower in glass than it is in water? [1]

Figure 8.21

4 Figure 8.22 shows how a lens can produce an image of an object. Each small square corresponds to a distance of 2 mm.
   (a) Which three words from the list below best describe the image? [3]
      real   diminished   erect   inverted   virtual   enlarged
   (b) What type of lens is shown in Figure 8.22? [1]

➔

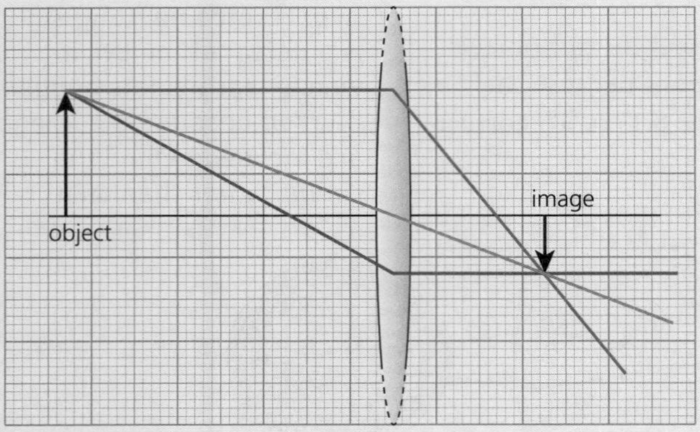

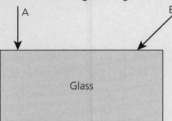

Figure 8.22

(c) Copy the diagram and mark the location of the principal focus on each side of the lens, and the optical centre. [3]

(d) The diagram is drawn to scale. Find the focal length of the lens in cm. [1]

(e) The magnification of the image is defined by the equation:

$$\text{magnification} = \frac{\text{height of image}}{\text{height of object}}$$

Use the equation to calculate the magnification of the image. [2]

5 A student was investigating how a ray of light passed through a rectangular glass block. He drew the diagram in Figure 8.23. Copy Figure 8.23 and show the correct paths of the two rays A and B through the glass and back into the air. [4]

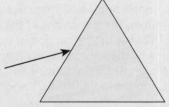

Figure 8.23

6 Figure 8.24 shows a ray of white light incident on a glass prism. When the light passes through the prism it splits into many different colours.

(a) What is this effect called? [1]

(b) State briefly why it happens. [2]

(c) Copy Figure 8.24 and show the passage of the coloured rays through the prism and into the air. [3]

Figure 8.24

F 7 A patient cannot see clearly an object that is about 25 cm from his eye. This is because the rays are brought to a virtual focus behind the retina (Figure 8.25).

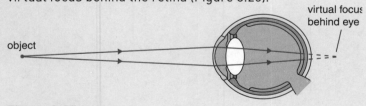

Figure 8.25

(a) State the medical name for the condition from which this patient is suffering. [1]

(b) What is the common name for this condition? [1]

(c) The patient has no glasses. What would she be inclined to do when reading a book? [1]

(d) What type of lens would an optician prescribe to correct for this problem in vision? [1]

(e) Draw a simple ray diagram to show how this type of lens corrects this problem in vision. [3]

Answers online

ONLINE

9 Electricity

We now know that there are two types of charge, **positive charge** and **negative charge**. The negative charge is due to the presence of electrons.

When we connect a battery across a lamp the lamp lights up (Figure 9.1). The connecting wire (copper) and the filament of the bulb (tungsten) are both **electrical conductors**. But the plastic covering is an **insulator**.

In general, *all* **metals** are electrical conductors. Almost all **non-metals** are insulators, but there are a few exceptions. For example, graphite is a non-metal, but it conducts electricity (Table 9.1).

> **electrical conductor** is a material through which electrical current passes easily
>
> **insulator** is a material through which electrical current cannot pass

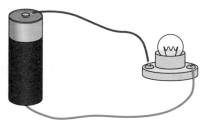

Figure 9.1 The lamp lights up when connected across a battery

Table 9.1 Common conductors and insulators

Good conductors	Gold	Silver	Copper	Aluminium	Mercury	Platinum	Graphite
Insulators	Polythene	Rubber	Wool	Wax	Glass	Paper	Wood

Why are metals good conductors? REVISED ☐

An electric current is a flow of electrically charged particles. At GCSE, the charge involved is always the electron. In metals, the outermost electron is often so weakly held that it can break away. We call such electrons **free electrons**. Some books call them delocalised electrons. However, in insulators there are *no* (or almost no) free electrons.

An electric cell (commonly called a battery) can make electrons move — but only if there is a conductor connecting its two terminals, making a complete circuit.

Scientists in the nineteenth century thought that an electric current consisted of a flow of positive charge from the positive terminal of the cell to the negative terminal. Unfortunately, although this idea is now known to be incorrect, this is still known as the direction of **conventional current**.

> **free electrons** are electrons that are not attached to any particular atom
>
> **conventional current** is the imagined current flowing from the positive terminal of a battery to the negative terminal

Summary

- Electrical conductors, like metals, have free electrons.
- Electrical insulators, like non-metals, have no free electrons.
- **Free electrons** move from the *negative* terminal to the *positive* terminal of the battery.
- **Conventional current** is said to flow from the *positive* terminal to the *negative* terminal of the battery.

Standard symbols

An electrical circuit may be represented by a **circuit diagram** with **standard symbols** for components (see Table 9.2).

> **standard symbols** are the internationally recognised symbols for electrical components

Table 9.2 Components and their symbols

Component	Symbol	Appearance
Switch		
Cell		
Battery		
Resistor		
Variable resistor		
Fuse	5A	
Voltmeter		
Ammeter		
Lamp		

Cell polarity

By convention, the long, thin line in the symbol for a cell is taken as the positive terminal. The short, fat line is the negative terminal.

Cells can be joined together minus–to–plus to make a battery. Cells connected in this way are said to be connected in **series**.

Connecting cells in series to make a battery increases the **voltage**. For example, connecting 4 × 1.5 volt cells in this way gives a 6 volt battery as shown in Figure 9.2.

> **cell polarity** is concerned with which end of a cell is positive and which end is negative
>
> a **series** circuit is one in which the components are arranged one-after-another, like carriages in a train
>
> **voltage** is the difference in electrical potential between two points that causes a current to flow

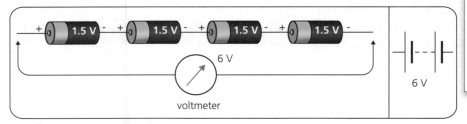

Figure 9.2 Cells correctly joined in series

If the polarity of one of the cells is reversed then the voltage is reduced dramatically. In Figure 9.3, the effect of reversing the polarity of the cell in the middle produces a battery of only 1.5 volts. The cells joined + to + cancel each other out.

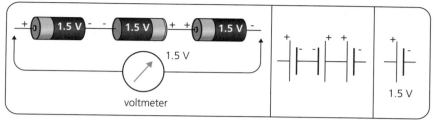

Figure 9.3 Cells incorrectly joined together

The relationship between charge and current

REVISED

The unit of charge is the **coulomb** (C).

The unit of **current** is the **ampere** (A). Currents of around 1 ampere and more can be measured by connecting an ammeter in the circuit. For smaller currents, a **milliammeter** is used. The unit in this case is the **milliampere** (mA) (1000 mA = 1 A). An even smaller unit of current is the **microampere** (μA) (1 000 000 μA = 1 A).

> **current** is a flow of electric charge in a circuit

In general, if a steady current I amperes flows for time t seconds, the charge Q coulombs passing any point is given by:

$$Q = I \times t$$

charge = current × time

(C) = (A) × (s)

Notice that:
- charge is measured in coulombs (C) and given the symbol Q
- current is measured in amperes (A) and given the symbol I
- time is measured in seconds (s) and given the symbol t.

So, a current of 1 ampere is flowing in a circuit if a charge of 1 coulomb passes a fixed point in 1 second.

Example

A current of 150 mA flows around a circuit for 1 minute. How much electrical charge flows past a point in the circuit in this time?

Answer

I = 150 mA = 0.15 A

t = 1 minute = 60 s

$Q = I \times t = 0.15 \times 60 = 9\,C$

Exam tip

Always ensure the units for current, charge and time are in amperes, coulombs and seconds before substituting values into the equation $Q = It$.

There are two conditions that *must* be met before an electric current will flow:
- There must be a *complete circuit* — i.e. there must be no gaps in the circuit.
- There must be a *source of energy* so that the charge may move — this source of energy may be a cell, a battery or the mains power supply.

Now test yourself

1 The cells in Figure 9.4 are all identical. The total battery voltage is 1.6 V.
  (a) Calculate the voltage of each cell.
  (b) State the maximum voltage that this battery could deliver.
2 Convert the following currents into milliamperes:
  (a) 3.0 A
  (b) 0.2 A
  (c) 200 µA
3 What charge is delivered if:
  (a) a current of 6 A flows for 10 seconds
  (b) a current of 300 mA flows for 1 minute
  (c) a current of 500 µA flows for 1 hour?

Figure 9.4

Answers online

Resistance

The **resistance**, R of an electrical conductor can be found using the ammeter–voltmeter method. We measure the current, I through the conductor when a voltage, V is applied across its ends. The resistance, R is then calculated using the equation:

$$R = \frac{V}{I}$$

where R is the resistance in ohms (Ω)
V is the voltage in volts
I is the current in amperes.

This is the basis of the Ohm's law experiment that follows.

> **resistance** is the opposition to the flow of current and is defined as the ratio of voltage to current

Prescribed practical P7

Ohm's law

Apparatus

- low-voltage power supply unit (PSU)
- rheostat
- ammeter
- voltmeter
- connecting leads
- resistance wire
- switch

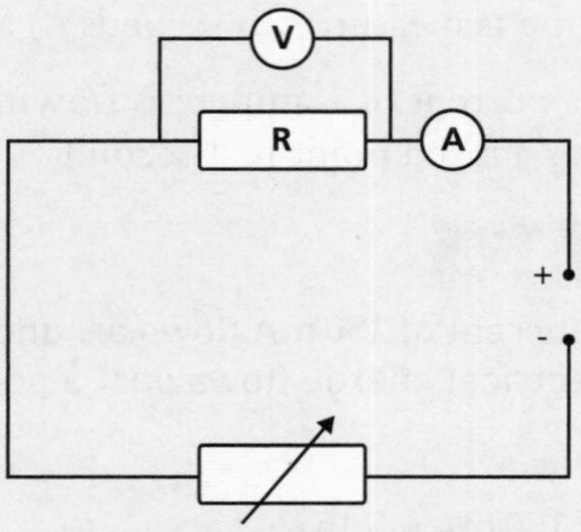

Method

1 Prepare a table for your results (see page 91).
2 Set up the circuit as shown in Figure 9.5.
3 Adjust the PSU to supply zero volts and then switch it on.
4 Record the voltage on the voltmeter and the corresponding current on the ammeter.
5 Record values for voltage and current, switch off the PSU and allow the wire to cool to room temperature.
6 Switch on the PSU and adjust the voltage so that the reading on the voltmeter increases by 0.5 V.

Figure 9.5 Circuit diagram

→

7 Repeat steps 5 and 6 for voltages from zero to a maximum voltage of 6 V*. This is Trial 1.
8 Repeat the entire experiment to obtain a second set of current values. This is Trial 2.
9 Calculate the mean current from the two trials.
10 Plot the graph of voltage against mean current.

*It is necessary to ensure the wire's temperature remains constant (close to room temperature). We do this by:
● keeping the voltage low (so that the current remains small), and
● switching off the current between readings to allow the wire to cool.

Table for results

Voltage/V	0.00	1.00	2.00	3.00	4.00	5.00	6.00
(Trial 1) current/A							
(Trial 2) current/A							
Mean current/A							
Ratio of voltage to current/Ω							

Treatment of the results

Plot the graph of voltage/V (y-axis) against mean current/A (x-axis) (Figure 9.6).

Evaluation of the results

The graph of V against I is a straight line through the origin. This tells us that the current in a metallic conductor is **directly proportional** to the voltage across its ends provided the temperature remains constant.

The last statement is commonly called **Ohm's law**.

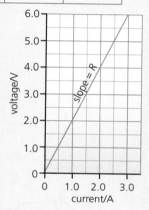

Figure 9.6 A graph showing voltage against current

---

direct proportion is the mathematical relationship between quantities that increase together in the same ratio. For example, when one quantity doubles the other quantity doubles also

Ohm's law states that the current in a metallic conductor is directly proportional to the voltage across its ends provided the temperature remains constant

Measuring the resistance

The resistance of a wire at constant temperature depends only on three factors:
● the material from which the wire is made
● the length of the wire, and
● the cross-sectional area of the wire.

The ratio $V{:}I$ is constant throughout the experiment.

Examples

1

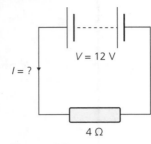

Figure 9.7

Answer

$V = 12\,V$, $R = 4\,\Omega$, $I = ?\,A$

current, $I = \dfrac{V}{R}$

$= \dfrac{12}{4}$

$= 3\,A$

2

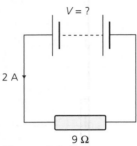

Figure 9.8

Answer

$I = 2\,A$, $R = 9\,\Omega$, $V = ?\,V$

voltage, $V = I \times R$

$= 2 \times 9$

$= 18\,V$

3

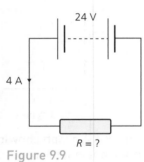

Figure 9.9

Answer

$V = 24\,V$, $I = 4\,A$, $R = ?$

resistance, $R = \dfrac{V}{I}$

$= \dfrac{24}{4}$

$= 6\,\Omega$

Now test yourself

4 Calculate the current flowing through a 10 Ω resistor which has a voltage of 20 V across it.
5 A resistor has a voltage of 15 V across it when a current of 3 A flows through it. Calculate the resistance of the resistor.
6 A current of 2 A flows through a 25 Ω resistor. Find the voltage across the resistor.
7 A voltage of 15 V is needed to make a current of 2.5 A flow through a wire.
   (a) What is the resistance of the wire?
   (b) What voltage is needed to make a current of 2.0 A flow through the wire?
8 There is a voltage of 6.0 V across the ends of a wire of resistance 12 Ω.
   (a) What is the current in the wire?
   (b) What voltage is needed to make a current of 1.5 A flow through the wire?
9 A resistor has a voltage of 6 V applied across it and the current flowing through it is 100 mA. Calculate the resistance of the resistor.
10 A current of 600 mA flows through a metal wire when the voltage across its ends is 3 V. What current flows through the same wire when the voltage across its ends is 2.5 V?

Answers online

Answers at **www.hoddereducation.co.uk/myrevisionnotesdownloads**

Practical

Resistance of a filament lamp

Variables

The independent variable is the current flowing in the lamp.
The dependent variable is the voltage across the lamp.
The controlled variables are the length of the filament (inside the lamp) and its area of cross-section.

Apparatus

- low-voltage power supply unit (PSU)
- rheostat
- ammeter
- voltmeter
- connecting leads
- filament lamp in a suitable holder
- switch

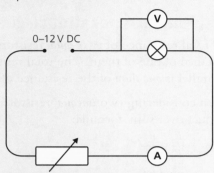

Figure 9.10 Circuit diagram

Method

1 Prepare a table for your results, as below.
2 Ensure the PSU is switched off and set up the circuit as shown in Figure 9.10.
3 Adjust the PSU to supply zero volts and switch it on.
4 Record the voltage on the voltmeter and current on the ammeter.
5 Adjust the voltage so that the voltmeter reading increases by 1.0 V.
6 Repeat steps 4 and 5 until readings have been recorded for voltages up to 6 V. This is Trial 1.
7 Repeat the entire experiment to obtain a second set of current values. This is Trial 2.
8 Calculate the mean current from the two trials.
9 Plot the graph of voltage against mean current.

Table for results

Voltage/V	0.00	1.00	2.00	3.00	4.00	5.00	6.00
(Trial 1) current/A							
(Trial 2) current/A							
Mean current/A							

Treatment of the results

Plot the graph of voltage/V (y-axis) against mean current/A (x-axis) (Figure 9.11).

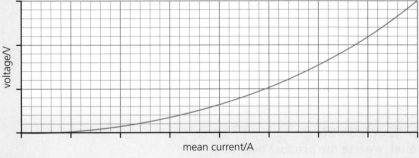

Figure 9.11

For a filament lamp, the graph of V against I is a curve with increasing gradient. This tells us that as the current in a lamp increases, the resistance of its filament also increases in a non-linear way.

Explanation

The resistance of a metal rises when the temperature increases. In this experiment the temperature of the filament is *allowed* to rise until it becomes white hot at its operating temperature.

Final task

You should now find the resistance of the filament at two points on your curve by calculating the ratio of the voltage to the current for the two different values of the current.

Resistance in series circuits

The total resistance of two or more resistors in series is simply the sum of the individual resistances of the resistors (Figure 9.12):

$$R_{total} = R_1 + R_2 + R_3$$

In Figure 9.12, the three resistors could be replaced by a single resistor of $(4 + 8 + 6) = 18\,\Omega$.

Resistance in parallel circuits

The total resistance of two *equal* resistors in **parallel** is *half* of the resistance of one of them. The total resistance of three *equal* resistors in parallel is *one third* of the resistance of one of them, and so on.

When considering two *unequal* resistors, R_1 and R_2, in parallel, we use the 'product over sum' formula:

$$R_{total} = \frac{R_1 \times R_2}{R_1 + R_2}$$

$$= \frac{\text{product}}{\text{sum}}$$

H The formula for calculating the total resistance of three resistors in parallel is:

$$\frac{1}{R_{total}} = \frac{1}{R_1} + \frac{1}{R_2} + \frac{1}{R_3}$$

This last formula can be extended to any number of resistors in parallel.

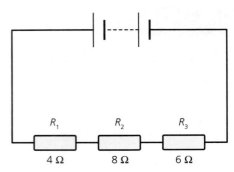

Figure 9.12 Calculating the total resistance of three resistors in a series circuit

a **parallel** circuit is one in which the current divides to travel independently along two or more separate loops

Examples

1 Find the combined resistance of two $8\,\Omega$ resistors:
   (a) in series
   (b) in parallel.

Answer

(a) For resistors in series: $R_{total} = R_1 + R_2 = 8 + 8 = 16\,\Omega$
(b) The total resistance of two *equal* resistors in parallel is *half* of the resistance of one of them.
   So in this case the total resistance is $\frac{8}{2} = 4\,\Omega$.

2 A $6\,\Omega$ resistor and a $3\,\Omega$ resistor are connected
   (a) in series and
   (b) in parallel.
      In each case find the resistance of the combination.

Answer

(a) For resistors in series: $R_{total} = R_1 + R_2 = 6 + 3 = 9\,\Omega$
(b) For two *unequal* resistors in parallel, we use the product over sum rule. So,

$$R_{total} = \frac{\text{product}}{\text{sum}} = \frac{(R_1 \times R_2)}{(R_1 + R_2)}$$

$$= \frac{(6 \times 3)}{(6 + 3)} = \frac{18}{9} = 2\,\Omega$$

→

Answers at **www.hoddereducation.co.uk/myrevisionnotesdownloads**

3 A 24 Ω resistor, a 12 Ω resistor and an 8 Ω resistor are connected
   (a) in series and
   (b) in parallel.
   In each case find the resistance of the combination.

Answer

(a) For resistors in series: $R_{total} = R_1 + R_2 + R_3 = 24 + 12 + 8 = 44\,\Omega$

(b) For three *unequal* resistors in parallel, we use:

$$\frac{1}{R_{total}} = \frac{1}{R_1} + \frac{1}{R_2} + \frac{1}{R_3}$$

$$= \frac{1}{24} + \frac{1}{12} + \frac{1}{8} = \frac{1}{24} + \frac{2}{24} + \frac{3}{24}$$

$$= \frac{6}{24}$$

$$= \frac{1}{4}$$

(Important: do not forget to turn the fraction upside down in the last step!)

So, $R_{total} = \frac{4}{1} = 4\,\Omega$

The final example illustrates how to cope with a combination of resistors in parallel with resistors in series.

4 Find the total resistance of the combination shown in Figure 9.13.

Figure 9.13

Answer

Using the product over sum formula, we see that the 4 Ω and 6 Ω parallel combination gives a total resistance of 2.4 Ω. Similarly, the 9 Ω and 18 Ω parallel combination gives a total resistance of 6.0 Ω.

There is therefore a series combination of 2.4 Ω + 1.6 Ω + 6.0 Ω, which gives a total resistance of 10 Ω.

Voltage in series circuits

REVISED

When resistors are connected in series:
● the current in each resistor is the same

Example

Resistances of 2 Ω, 4 Ω and 6 Ω are connected in series across a 3 V battery.
(a) Calculate:
   (i) the total resistance of the circuit
   (ii) the current in each resistor
   (iii) the voltage across each resistor.
(b) Comment on the answers to part (a)(iii).

Answer

First, draw the circuit diagram (Figure 9.14).

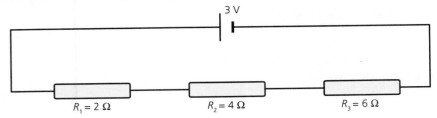

Figure 9.14

(a) (i) $R_{total} = R_1 + R_2 + R_3$

$= 2 + 4 + 6$

$= 12\,\Omega$

This means that the circuit is equivalent to one with $12\,\Omega$ placed across a 3V battery.

(ii) $I = \dfrac{V}{R} = \dfrac{3}{12} = 0.25\,A$

(iii) $2\,\Omega: V = I \times R = 0.25 \times 2 = 0.5\,V$

$4\,\Omega: V = I \times R = 0.25 \times 4 = 1.0\,V$

$6\,\Omega: V = I \times R = 0.25 \times 6 = 1.5\,V$

(b) The sum of the voltages (0.5V + 1.0V + 1.5V) is 3V, which is exactly the same as the voltage of the battery.

The voltages are in exactly the *same proportion* as the resistances. So, the $4\,\Omega$ resistor has twice the voltage as the $2\,\Omega$ resistor and the $6\,\Omega$ resistor has three times the voltage as the $2\,\Omega$ resistor.

● the sum of the voltages across each resistor is equal to the battery voltage.

Voltage in parallel circuits

REVISED

When resistors are connected in parallel:

Examples

1 Resistances of $2\,\Omega$ and $3\,\Omega$ are connected in parallel across a 6V battery.
   (a) State the voltage across reach resistor.
   (b) Calculate:
      (i)   the total resistance of the circuit
      (iii) the current in each resistor
      (iii) the total current taken from the battery.

Answer

(a) 6V

(b) (i) R_{total} = product over sum

$$= \dfrac{(R_1 \times R_2)}{(R_1 + R_2)} = \dfrac{(2 \times 3)}{(2 + 3)} = \dfrac{6}{5} = 1.2\,\Omega$$

(ii) $2\,\Omega: I = \dfrac{V}{R} = \dfrac{6}{2} = 3\,A$

$3\,\Omega: I = \dfrac{V}{R} = \dfrac{6}{3} = 2\,A$

→

(iii) We can do this in two different ways:

$$I_{battery} = \frac{V_{battery}}{R_{total}} = \frac{6}{1.2} = 5\,A$$

or $I_{battery}$ = sum of currents in resistors
$$= 3 + 2$$
$$= 5\,A$$

2 Resistances of $4\,\Omega$, $6\,\Omega$ and $12\,\Omega$ are connected in parallel across a battery. A current of 2.0 A flows from the battery towards the parallel network.

(a) Calculate:
  (i)   the total resistance of the network
  (ii) the battery voltage and the voltage across each resistor
  (iii) the current in each resistor.

(b) Comment on the answers to parts (a)(ii) and (iii).

Answer

First, draw the circuit diagram (Figure 9.15).

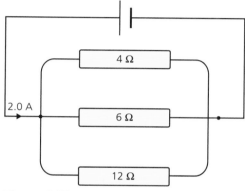

Figure 9.15

(a) (i)
$$\frac{1}{R_{total}} = \frac{1}{R_1} + \frac{1}{R_2} + \frac{1}{R_3}$$
$$= \frac{1}{12} + \frac{1}{6} + \frac{1}{4} = \frac{6}{12} = \frac{1}{2}$$

So, $R_{total} = \frac{2}{1} = 2\,\Omega$

(ii) $V_{battery} = I_{battery} \times R_{total}$
$$= 2.0 \times 2 = 4\,V$$
voltage across each resistor = 4 V

(iii) $4\,\Omega$: $I = \frac{V}{R} = \frac{4}{4} = 1.0\,A$

$6\,\Omega$: $I = \frac{V}{R} = \frac{4}{6} = 0.67\,A$

$12\,\Omega$: $I = \frac{V}{R} = \frac{4}{12} = 0.33\,A$

(b) In (a)(ii) the voltages across each resistor are the same as the battery voltage.

In (a)(iii) the sum of the currents in the parallel network (1.0 A + 0.67 A + 0.33 A) is 2.0 A, which is exactly the same as the current from the battery.

The currents in each resistor are in **inverse proportion** to the resistance. So the current in the $6\,\Omega$ resistor is twice the current in the $12\,\Omega$ resistor, and the current in the $4\,\Omega$ resistor is 3 times the current in the $12\,\Omega$ resistor.

Typical mistake

The most common error when using the 'one-over-R' formula for resistors in parallel is forgetting to flip it at the end of the calculation. So remember — don't forget to flip!

inverse proportion is the mathematical relationship between quantities in which one quantity doubles when the other halves

- the voltage across each resistor is the same as the voltage provided by the battery
- the sum of the currents in each resistor is equal to the current coming from the battery.

Hybrid circuits

Hybrid circuits contain both series and parallel elements. We treat each part separately, eventually finding the total resistance of the entire network as shown in the example.

Example

A parallel combination of 2 Ω and 3 Ω is joined in series with a 5 Ω resistor (Figure 9.16). The network is connected across a 31 V battery. Calculate the current taken from the battery, the voltage across each resistor and the current in each resistor.

(The strange battery voltage has been chosen to make the arithmetic easy.)

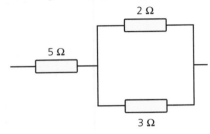

Figure 9.16

Answer

From Example 1 on p. 96 we see that the combined resistance of the two resistors in parallel is 1.2 Ω.

So the total resistance of the circuit, $R_{total} = 5 + 1.2 = 6.2\,\Omega$

The current taken from the battery $I_{battery} = \dfrac{V_{battery}}{R_{total}}$

$$= \dfrac{31}{6.2} = 5\,A$$

The current in the 5 Ω resistor is therefore 5 A.

The voltage across the 5 Ω resistor is $V = I \times R$

$$= 5 \times 5 = 25\,V$$

Since the battery voltage is 31 V, the voltage across the parallel network is 31 − 25 = 6 V

So the voltage across the 2 Ω resistor = voltage across the 3 Ω resistor = 6 V

Current in 2 Ω resistor $= \dfrac{V}{R} = \dfrac{6}{2} = 3\,A$

and current in 3 Ω resistor $= \dfrac{V}{R} = \dfrac{6}{3} = 2\,A$

Exam tip

When doing maths questions in physics remember the examiner wants to see:
- your equation or formula
- your substitutions
- your arithmetic
- your answer
- your unit.

Answers at **www.hoddereducation.co.uk/myrevisionnotesdownloads**

Now test yourself

11 (a) Calculate the value of the current from the cell in each of the circuits in Figure 9.17.

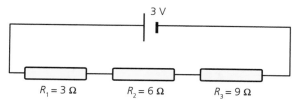

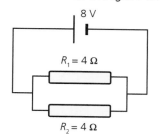

Figure 9.17

(b) State the voltage across each 4Ω resistor in the second circuit.

12 Two identical resistors are placed in parallel across a 12 V battery. If the total current drawn from the battery is 3 A, what is the resistance of each resistor?

Answers online

Prescribed practical P8

Resistance and length

Variables

The independent variable is the length of the wire.

The dependent variable is the resistance of the wire.

The controlled variables are the temperature and the area of cross-section of the wire.

Apparatus

- low-voltage power supply unit (PSU)
- rheostat
- ammeter
- voltmeter
- connecting leads
- resistance wire
- switch
- metre ruler

Method

1 Prepare a table for your results (see page 100).
2 Measure and cut off 1 metre of nichrome resistance wire.
3 Attach it with sticky tape to a metre ruler and set up the circuit as shown in Figure 9.18.
4 Ensure that the PSU is switched off.
5 Adjust the PSU to supply about 1 volt.
6 Connect the 'flying lead' so that the length of wire across the voltmeter is 10 cm.
7 Switch on the PSU and record the voltage on the voltmeter and the current on the ammeter.

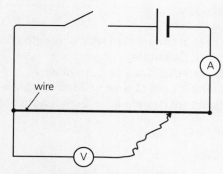

Figure 9.18 Circuit diagram

→

8 Switch off the PSU immediately and allow the wire to cool to room temperature*.

9 Increase the length across the voltmeter by 10 cm.

10 Repeat steps 7 to 9 until readings have been recorded for lengths from 10 cm to 90 cm.

11 Calculate the resistance of each length of wire, using $R = \dfrac{V}{I}$.

12 Plot the graph of resistance (y-axis) against length (x-axis).

*It is necessary to ensure the wire's temperature remains constant (close to room temperature). We do this by:
- keeping the voltage low (so that the current remains small) and
- switching off the current between readings to allow the wire to cool.

Table for results

Length of wire/cm	10	20	30	40	50	60	70	80	90
Voltage across wire/V									
Current in wire/A									
Resistance of wire/Ω									

Treatment of the results

Plot the graph of resistance/Ω (y-axis) against length/cm (x-axis).

Evaluation of the results

The graph of resistance/Ω against length/cm is a straight line through the origin (Figure 9.19). This tells us that the resistance of a metal wire is **directly proportional** to its length, provided its area of cross-section and the temperature remain constant.

The relationship between R and L is:

$R = kL$

where k is the gradient of the graph.

Since $k = \dfrac{R}{L}$, the unit for k is Ω/cm (or Ω/m).

Note that the value of k depends on the material of the wire and its cross-sectional area.

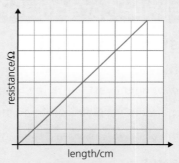

Figure 9.19 Graph of resistance against cross-sectional area

Example

A reel of constantan wire of length 250 cm has a total resistance of 15.0 Ω. Calculate:

(a) the resistance of 1.0 m of wire

(b) the length of wire needed to have a resistance of 3 Ω

(c) the resistance of a 90 cm length of the wire.

Answer

(a) $k = \dfrac{R}{L}$

$= \dfrac{15\,\Omega}{250\,\text{cm}} = 0.06\,\Omega/\text{cm} = 6\,\Omega/\text{m}$

so the resistance of 1.0 m of wire is 6 Ω

(b) $L = \dfrac{R}{k}$

$= \dfrac{3\,\Omega}{0.06\,\Omega/\text{cm}} = 50\,\text{cm}$

(c) $R = k \times L$

$= 6\,\Omega/\text{m} \times 0.9\,\text{m} = 5.4\,\Omega$

Practical

Resistance and cross-sectional area

Variables

The dependent variable is the resistance of each wire.

The independent variable is the cross-sectional area of each wire.

The controlled variables are the length of the wire and the material from which it is made.

Apparatus

- low-voltage power supply unit (PSU)
- rheostat
- ammeter
- voltmeter
- connecting leads

- filament lamp in a suitable holder
- switch
- wooden dowel or pencil
- micrometer

Method

1 Prepare seven samples of constantan wire, all 50 cm long and all of different cross-sectional area.
2 Prepare a table for your results like that shown below.
3 As a preliminary, use the micrometer screw gauge to measure the diameter, d, of one of the wires or, more simply, measure the length (l) of 20 turns of a resistance wire wound tightly together on a pencil or wooden dowel. Divide this length by 20 to calculate its diameter.
4 Calculate the cross-sectional area, A, using $A = \frac{\pi d^2}{4}$ and record the data in the table of results.
5 Repeat this process for six further thicknesses of the same length of wire and same type of material.
6 Ensure that the PSU is switched off and connect it to the mains socket.
7 Set up the circuit as shown in Figure 9.20.

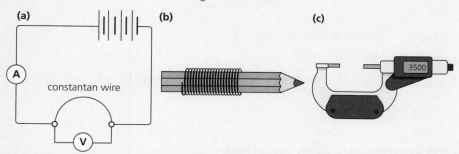

(a)   **(b)**   **(c)**

constantan wire

Figure 9.20 (a) Circuit diagram (b) Wire wrapped around pencil (c) Micrometer

8 Switch on the PSU and adjust it as necessary to obtain a voltage of 2 V.
9 Record the voltage on the voltmeter and the corresponding current on the ammeter.
10 Determine the resistance of this specimen of wire using $R = \frac{V}{I}$ and record the data in the table.
11 Repeat steps 7 to 10 for each of the other wire specimens.
12 Plot the graph of resistance against cross-sectional area and resistance against $\frac{1}{\text{area}}$.

Table for results

Mean diameter/mm							
Cross-sectional area/mm²							
Voltage/V							
Current/A							
Resistance/Ω							
(1/area)/(1/mm²)							

H Treatment of the results

Plot the graphs of:

(i) resistance/Ω against cross-sectional area/mm^2 (Figure 9.21)

(ii) resistance/Ω against $\dfrac{1/\text{area}}{1/\text{mm}^2}$ (Figure 9.22).

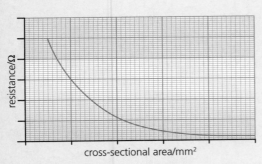

Figure 9.21 Graph of resistance against cross-sectional area

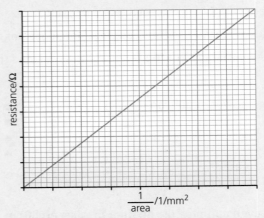

Figure 9.22 Graph of resistance against 1/area

Evaluation of the results

The graph of resistance R against cross-sectional area A is a curve with decreasing gradient.

The graph of R against $1/A$ is a straight line through the origin. This tells us that the resistance is inversely proportional to the cross-sectional area.

This means the mathematical relationship between the resistance, R and the cross-sectional area, A is $R = \dfrac{k}{A}$

where k is the gradient of the straight line graph of R against $1/A$.

Since $k = RA$, the unit for k is $\Omega\,\text{cm}^2$ (or $\Omega\,\text{m}^2$ or $\Omega\,\text{mm}^2$).

Note that, for a given wire, the value of k depends on the material of the wire and its length.

Example

A length of constantan wire has a resistance of $15.0\,\Omega$ and a cross-sectional area of $0.3\,\text{mm}^2$. Calculate:

(a) the resistance of the same length of constantan wire of cross-sectional area $0.9\,\text{mm}^2$

(b) the cross-sectional area of the same length of constantan wire if its resistance is $9\,\Omega$.

Answer

(a) $k = RA = 15\,\Omega \times 0.3\,\text{mm}^2 = 4.5\,\Omega\,\text{mm}^2$

$R = \dfrac{k}{A} = \dfrac{4.5\,\Omega\,\text{mm}^2}{0.9\,\text{mm}^2} = 5\,\Omega$

(b) $A = \dfrac{k}{R} = \dfrac{4.5\,\Omega\,\text{mm}^2}{9\,\Omega} = 0.5\,\text{mm}^2$

Answers at www.hoddereducation.co.uk/myrevisionnotesdownloads

Practical

Investigating the resistance of a metallic conductor at constant temperature

The resistance of metal conductors at constant temperature depends on the material from which the conductor is made.

The controlled variables are the length and the thickness of wire.

Using 1 metre of 32 swg copper wire, measure and record the resistance as before.

Then repeat the process using the same dimensions of wires such as manganin, nichrome and constantan.

When comparing wires of the same length and cross-sectional area, you should find that the order of increasing resistance is: copper, manganin, constantan and nichrome.

Now test yourself

TESTED ☐

13 A school buys a reel of constantan wire. The supplier's data sheet says that the wire has a resistance of 2.5 Ω/m. Calculate:
   (a) the length of wire a technician must cut from the reel to give a resistance of 2 Ω
   (b) the resistance of a 120 cm length of wire cut from the reel.

14 A technician cuts an 80 cm length of wire from a reel marked 3.0 Ω/m.
   The technician joins the two free ends of the wire together to form a loop. She then attaches two crocodile clips to the wire at opposite ends of a diameter.
   (a) Explain why the total resistance between the crocodile clips is 0.6 Ω.
   (b) In what way, if at all, does the total resistance between the crocodile clips change if one of the clips is moved along the wire towards the other. Explain your reasoning.

15 A 50 cm length of wire of diameter 0.2 mm has a resistance of 1.6 Ω. Find the resistance of:
   (a) a 75 cm length of wire of the same material and same diameter
   (b) a 50 cm length of wire of the same material and diameter 0.4 mm
   (c) a 75 cm length of wire of the same material and diameter 0.4 cm.

Answers online

Exam practice

1 (a) What flows in the direction indicated by the arrow in Figure 9.23? [1]

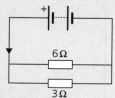

Figure 9.23

(b) Copy Figure 9.23 and mark on it an arrow to show the direction in which charged particles flow through the two resistors. [1]
(c) What name is given to these charged particles? [1]
(d) The current taken from the battery is 0.6 A. State the size of the current in the smaller resistor. [1]
(e) Show that the electrical charge delivered by the battery in 1 minute is 36 C. [3]

→

2 Four identical resistors are arranged as shown in Figure 9.24.

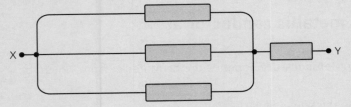

Figure 9.24

The current entering at X is 3 mA and the voltage across the isolated series resistor is 9 mV. Calculate:

(a) the total resistance between X and Y [3]

(b) the resistance of each resistor [2]

(c) the current in each resistor [2]

(d) the voltage across each resistor [2]

3 In Figure 9.25 resistors R_1 and R_2 have resistances of 80 Ω and 40 Ω respectively.

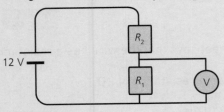

Figure 9.25

(a) Calculate the voltage you would expect to observe on the voltmeter. [3]

(b) What assumption have you made about the resistance of the voltmeter itself? [1]

4 Copy and complete the table below to show the effective resistance between X and Y for different switch settings in Figure 9.26. All four resistors have a resistance of 6 Ω.

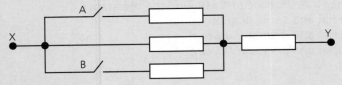

Figure 9.26

Switch		Effective resistance between X and Y/Ω
A	B	
Open	Open	
Open	Closed	
Closed	Open	
Closed	Closed	

Answers online

10 Electricity in the home

Electric current generates heat when it passes through a metal wire. Why does this happen?

As electrons pass through the conductor they collide with the atoms. In these collisions the light electrons lose energy and the heavy atoms gain energy. This causes the atoms to vibrate faster. Faster vibrations mean a higher temperature. This is called **joule heating**. Hairdryers and toasters (Figure 10.1) both use the joule heating effect of an electric current

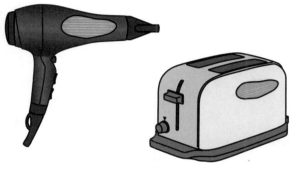

Figure 10.1 Toasters and hairdryers use an electrical current to produce heat

Electrical energy

If 1 coulomb of charge gains or loses 1 joule of energy between two points, there is a voltage of 1 volt between those two points.

$$\text{voltage} = \frac{\text{energy transferred}}{\text{charge}}$$

Rearranging this formula gives us:

$$\text{energy transferred} = \text{voltage} \times \text{charge}$$

or

$$\text{energy transferred} = \text{voltage} \times \text{current} \times \text{time}$$

$$E \quad = \quad V \quad \times \quad I \quad \times \quad t$$

Electrical power

Earlier you learned that the formula for mechanical power of a machine is defined as the rate at which energy is transferred and is given by:

$$\text{power} = \frac{\text{energy transferred}}{\text{time}}$$

We have seen that in an electrical circuit:

$$\text{energy transferred} = \text{voltage} \times \text{current} \times \text{time}$$

Substituting for energy transferred:

$$power = \frac{voltage \times current \times time}{time}$$

or

electrical power = voltage × current

> **electrical power** is the rate at which electrical energy is used in a circuit
>
> **Joule's law** states that the power generated in an electrical component is the product of the current in the component and the voltage across it ($P = IV$)

This is often expressed by the equation $P = I \times V$ and is commonly called **Joule's law** of heating.

Example

A study lamp is rated at 60 W, 240 V. How much current flows in the bulb?

Answer

$$power = V \times I$$

$$60 = 240 \times I$$

$$I = \frac{60}{240} = 0.25\,A$$

Now test yourself

1 Calculate the power of a heater if it uses 3 600 000 J in 1 hour.
2 A toaster has a power rating of 1.5 kW.
   (a) State the power of the toaster in watts.
   (b) How much energy does the toaster use in 10 s?
3 A filament bulb has an internal resistance of 960 Ω when it is at normal brightness.
   (a) Calculate the current it draws when connected to a 240 V supply and is shining with normal brightness.
   (b) Calculate its power at normal brightness.

Answers online

The three equations for power

REVISED

Because power $P = I \times V$ and Ohm's law is $V = I \times R$ we can write:

$$P = IV$$

$$P = I^2 R$$

$$P = \frac{V^2}{R}$$

These formulae give the electrical power **dissipated** (converted) into heat in resistors and heating elements. The heat dissipated is sometimes referred to as 'ohmic losses'.

Example

What power is dissipated in a 10 Ω resistor when the current through it is: (a) 2 A (b) 4 A?

Answer

(a) power = $I^2 R$ = $2^2 \times 10 = 40$ W

(b) power = $I^2 R$ = $4^2 \times 10 = 160$ W

The example shows that when the current is doubled, the power dissipated is quadrupled. This idea has important implications for electricity transmission in the next chapter.

Now test yourself

TESTED

4 A washing machine uses 1200 W when connected to a 240 V power supply.
   (a) Calculate the current that it draws in normal use.
   (b) Would the power of the machine increase, decrease or remain the same if it was connected to a 120 V supply?

5 Copy and complete the table for domestic appliances, all of which operate at 240 V.

Name of appliance	Power rating	Current drawn	Resistance
Filament lamp		0.25 A	
Coffee machine	120 W		
Iron		6 A	
Electric oven			24 Ω
Immersion heater	3 kW		

Answers online

Paying for electricity

REVISED

Electricity companies bill customers for electrical energy in units known as **kilowatt-hours** (kWh).

The following two formulae are very useful in calculating the cost of using a particular appliance for a given amount of time:

number of units used = power rating (in kilowatts) × time (in hours)

total cost = number of units used × cost per unit

> One **kilowatt-hour** is the amount of energy transferred when 1000 W is delivered for 1 hour.

Example

If electricity costs 16 pence per kWh, find the number of units used by a 3000 W immersion heater when it is switched on for $1\frac{1}{2}$ hours. Calculate also the cost of using this immersion heater for that time.

Answer

number of units used = power rating (in kilowatts) × time (in hours)

$$= 3\,kW × 1.5\,h = 4.5\,kWh$$

total cost = number of units used × cost per unit

$$= 4.5\,kWh × 16\,pence/kWh$$

$$= 72\,pence = £0.72$$

Exam tip

When doing questions on the cost of using electricity ask yourself if your answer is reasonable. If you get the cost of taking a shower to be £15 (which is unreasonable), you have almost certainly made a simple mistake like giving the wrong unit — 15 pence is much more likely than £15.

Electricity bills

Figure 10.2 shows part of a typical electricity bill. The difference between the present reading and the previous reading is the number of units used. In this particular example, the number of units used = 57 139 − 55 652 = 1487 units (kWh)

Northern Electricity Board				Customer account no: 3427 364
Present meter reading	Previous meter reading	Units used	Cost per unit (incl. VAT)	£
57139	55652	1487	15.0p	£223.05

Figure 10.2 An electricity bill

If the cost of a unit is known, then the total cost of the electricity used can be determined. In the example in Figure 10.2, 1487 units at 15.0 pence per unit = 22 305 pence = £223.05.

One-way switches

This kind of switch acts as a make-or-break device to switch a circuit on or off (Figure 10.3). When the switch is open there is air between the conducting contacts. Since air is an insulator the circuit is incomplete. No current flows.

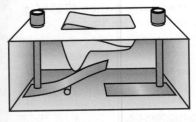 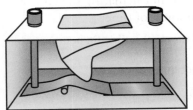

Figure 10.3 A one-way switch

The rocker itself is made of plastic. This is important with high voltages to prevent current flowing through the body of the user. When the rocker is pressed the conducting contacts are pushed together.

There is now a complete circuit, so current can flow.

As we shall see later, it is important that the switch is placed in the live side of a circuit.

Answers at www.hoddereducation.co.uk/myrevisionnotesdownloads

Two-way switches

In most two-storey houses, you can turn the landing lights on or off from upstairs or downstairs. Two-way switches are used for this.

Figure 10.4(a) illustrates a two-way switch in one of its two on-positions. Going to bed at night, when both switches are up (or both are down), the circuit is complete and a current flows through the bulb.

At the top of the stairs, one of the switches is pressed down and the circuit is broken (Figure 10.4b). You should verify for yourself what happens when the person comes downstairs. The effect is that each switch reverses the effect of the other one (Figure 10.5).

(a) **(b)**

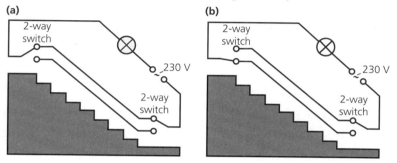

Figure 10.4 A two-way switch circuit

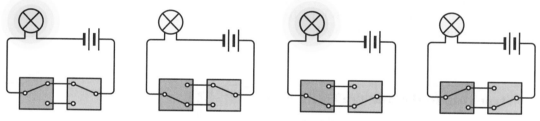

Figure 10.5 The four possible states of a two-way switch in a circuit

How to wire a three-pin plug

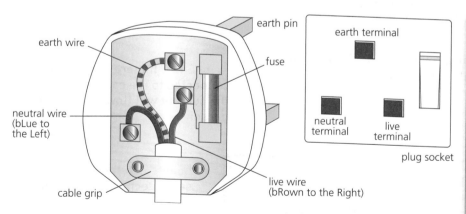

Figure 10.6 A correctly wired three-pin plug

the **neutral** wire is the wire in a mains voltage system that is permanently at zero volts

the **live** wire is the dangerous wire in a mains voltage system in which the voltage is changing

the **earth** wire is the wire that connects the metal frame in an electrical appliance to the earth

- The wire with the **blue** insulation is the **neutral** wire — connect this to the left-hand pin (Figure 10.6).
- The **brown** insulated wire is the **live** wire — connect this to the right-hand pin.
- The wire with the **yellow and green** insulation is the **earth** wire — connect this to the top pin.

Fuses

A **fuse** is a device that is meant to prevent damage **to an appliance**. The most common fuses are either a 3 A (red) fuse for appliances up to 720 W or 13 A (brown) fuse for appliances between 720 W and 3 kW.

If a larger-than-usual current flows, the fuse wire will melt and so break the circuit.

> a **fuse** is a short wire, often used inside a plug, which melts to disconnect a circuit when too much current flows

Selecting a fuse

Every appliance has a power rating. How much current the appliance will use is found using the power formula:

power = voltage × current

For example, a jig-saw has a power rating of 350 W. So the current it draws when connected to the mains is given by:

$$current = \frac{power}{voltage} = \frac{350}{240} = 1.46\,A$$

This is the normal current the device uses — any larger current can destroy it.

A 3 A fuse would allow a normal working current to flow and protect the jig-saw from larger currents. A 13 A fuse would allow a dangerously high current to flow and still not 'blow'.

So, it is important to use the correct size of fuse.

Remember that a fuse **protects the appliance**. It does not protect the person using the appliance. It can take 1 to 2 seconds for a fuse to melt — enough time for the user to receive a fatal electric shock.

The earth wire

REVISED

An **earth wire** can prevent harm **to the user** (see Figure 10.7).

Suppose a fault develops in an electric fire and the element is in contact with the metal casing of the fire. The casing will be live and if someone were to touch it they would get a possibly fatal electric shock as the current rushes through their body to earth. The earth wire prevents this — it offers a **low resistance** route of escape, enabling the current to go to earth by a wire rather than through a human body.

Any appliance with a **metal casing** could become live if a fault develops, so such appliances should always have a fuse fitted in the plug.

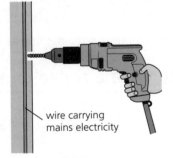

wire carrying mains electricity

Figure 10.7 If there is no earth wire connected to the casing of the drill, the current will flow through the person

Double insulation

REVISED

Appliances such as vacuum cleaners and hairdryers are usually **double insulated**. The appliance is encased in an insulating plastic case and is connected to the supply by a two-core insulated cable containing only a live and a neutral wire. Any metal attachments that a user might touch are fitted into the plastic case so that they do not make a direct connection with the motor or other internal electrical parts. The symbol for double insulated appliances is shown in Figure 10.8.

Figure 10.8 The symbol for double insulation

> **double insulation** is a safety technique in which an electrical appliance is encased in an insulating plastic case to prevent a user coming into contact with a live component and being harmed

Answers at www.hoddereducation.co.uk/myrevisionnotesdownloads

The live wire

Earlier we said that the switch and fuse must be placed on the live side of the appliance (Figure 10.9). Why is this important? The live wire in a mains supply is at high voltage (effectively around 230 V). The neutral side is at approximately zero volts.

If a fault occurs and the fuse blows, the live, dangerous wire is disconnected. If the fuse were on the neutral side, the appliance would still be live, even when the fuse had blown.

Switches are also placed on the live side for the same reason. If the switch were on the neutral side, the appliance would still be live, even when the switch was in the OFF position.

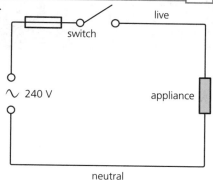

Figure 10.9 This fuse has been correctly placed on the live side of the power supply

Exam practice

1 (a) What is the purpose of the fuse in a three-pin plug?   [1]
   (b) State the colours of the live, neutral and earth wires in a three-pin plug.   [3]
   (c) What is the purpose of the earth wire?   [1]
   (d) Some appliances have only two wires in their plugs. Explain why.   [1]
2 A user of electricity receives a bill showing that he has used 816 units from 1 May until 1 August. Electricity costs 15 pence per unit.
   (a) What is the scientific name for the commercial unit of electrical energy?   [1]
   (b) Calculate the cost of the electricity used.   [2]
   (c) In the following quarter the cost rises from 15 pence to 16 pence per unit. What is the maximum number of units the customer can use if his bill is to be no greater than your answer to part (b)?   [2]
3 (a) Some appliances are double insulated. What does this mean?   [1]
   (b) Explain how double insulation can protect a user from electric shock.   [2]
   (c) A schoolboy incorrectly wires a three-pin plug by connecting the neutral wire to the fuse and the live wire to the neutral terminal. His teacher explains that although the appliance connected to the plug would still work, his mistake would make it very dangerous. Explain why the appliance would be dangerous.   [1]
4 A music centre has a maximum power rating of 125 W and is connected to a 240 V supply.
   (a) What fuse should be fitted inside its plug? Choose from: 1 A, 3 A, 5 A and 13 A.   [2]
   (b) How much would it cost to run the music centre continuously for 8 hours if electricity costs 16 pence per unit?   [3]

Answers online

11 Magnetism and electromagnetism

Magnetic field pattern around a bar magnet REVISED

Magnetic fields can be investigated using a small **plotting compass**. The 'needle' is a tiny magnet that is free to turn on its spindle.

Figure 11.1 shows how a plotting compass can be used to plot the field around a bar magnet.
- The **field lines** run from the north **pole** (N) to the south pole (S) of the magnet (Figure 11.2).
- The field direction, shown by the arrowhead, is defined as the direction in which the force on a N pole would act. This means that magnetic field lines must never touch or cross over each other.
- The magnetic field is strongest where the field lines are closest together, i.e. at the poles of the magnet. A point where there is no magnetic field is called a neutral point. (Figure 11.3)

> **magnetic field** is a region of space within which a magnet experiences a force
>
> **plotting compass** is a small compass used to find the shape of a magnetic field
>
> **magnetic field lines** are lines drawn to represent a magnetic field
>
> **poles** (of a magnet) are the ends of a magnet (north and south) where the magnetic field is strongest

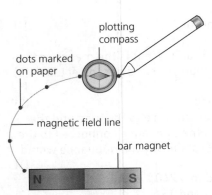

Figure 11.1 Plotting the field around a bar magnet using a compass

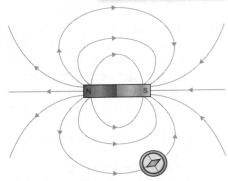

Figure 11.2 Magnetic field lines around a bar magnet

(a)

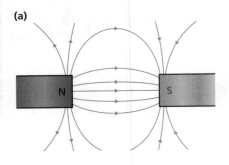

(b)

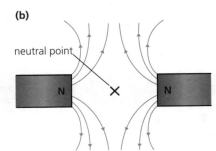

Figure 11.3 Field lines occurring when (a) a N and S pole are brought close together and (b) two N poles are brought close together

Magnetic field pattern due to a current-carrying coil REVISED

A coil with one turn

Figure 11.4 shows the magnetic field around a **single** loop of wire that is carrying a current.

Answers at **www.hoddereducation.co.uk/myrevisionnotesdownloads**

The strength of the magnetic field can be increased by:
- increasing the number of turns of wire in the coil
- increasing the current through the coil.

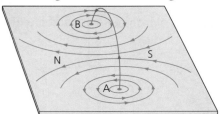

Figure 11.4 The magnetic field around a single loop of current-carrying wire

A coil with many turns

A stronger magnetic field can be made by wrapping a wire in the form of a long coil, which is referred to as a **solenoid**. A current must be passed through it, as illustrated in Figure 11.5.

The magnetic field produced by a solenoid has the following features:
- The field is similar to that from a bar magnet, with poles at the ends of the coils.
- Increasing the current increases the strength of the field.
- Increasing the number of turns on the coil increases the strength of the field.

Polarity

To work out which way round the poles are you can use the **right-hand grip rule**, as shown in Figure 11.6. Imagine gripping the solenoid with your right hand so that your fingers point in the conventional current direction. Your thumb points towards the north pole of the solenoid.

Now test yourself   TESTED ☐

1 Explain why magnetic field lines can never cross or touch.
2 Explain the meaning of the term *neutral point*.

Answers online

> **solenoid** is a coil of wire carrying an electrical current

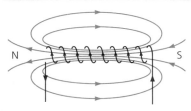

Figure 11.5 The pattern of magnetic field around a solenoid

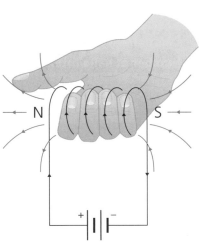

Figure 11.6 The right-hand grip rule

Prescribed practical P9

Factors affecting the strength of an electromagnet

The strength of an electromagnet, i.e. the strength of its magnetic field, can be measured by finding the mass of iron it will attract. Iron nails or paper clips may be used. Three factors are involved in this activity.

1 Investigating the effect of the current on the strength of the magnetic field

Apparatus
- thick insulated coil of copper wire
- soft iron core
- ammeter
- iron nails
- variable power supply

→

Method

1 Construct an electromagnet using, for example, 50 turns of insulated wire around a soft iron core.
2 Connect it to the circuit as shown in Figure 11.7.
3 Using the rheostat, increase the current in steps, measuring the number of iron nails attracted to the electromagnet for each current.
4 Record your results in a suitable table.
5 Plot a graph of number of nails on the y-axis versus current on the x-axis.

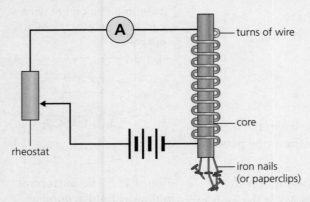

Figure 11.7 **Investigating the strength of an electromagnet**

Table for results

Current/A	0.0	0.5	1.0	1.5	2.0	2.5	3.0	3.5
Number of nails lifted								

A graph similar to the one shown in Figure 11.8 should be obtained. It shows that the number of nails lifted (and so the strength of the magnetic field) increases as the current increases. They are not directly proportional as the graph is not a straight line through the origin.

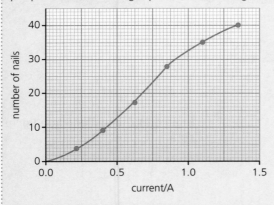

Figure 11.8 **A typical graph of results**

2 Investigating the effect of the number of turns on the strength of the magnetic field

Apparatus

- thick insulated coil of copper wire
- soft iron core
- ammeter
- iron nails
- variable power supply

Method

1 Keeping the current at 2.5 A and the material of the core constant, increase the number of turns of wire in steps of 10.
2 For each number of turns measure the number of nails lifted.

→

Answers at **www.hoddereducation.co.uk/myrevisionnotesdownloads**

3 Record your results in a suitable table.

4 Plot a graph of number of turns versus number of nails lifted.

Table for results

Number of turns	10	20	30	40	50	60
Number of nails lifted						

We would expect the number of nails lifted to increase as the number of turns increases.

3 Investigating the effect of the material of the core on the strength of the magnetic field

Apparatus

- thick insulated coil of copper wire
- cores made from various materials
- iron nails
- power supply

Method

1 This time, keep the number of turns of wire and the current constant at 3.0 A but change the material of the core from soft iron to steel to plastic, wood or no material.

2 In each case measure the number of iron nails lifted.

3 Record your results in a table.

4 Draw a **bar chart** of your results and describe your findings.

Table for results

Type of material	Number of nails lifted
Soft iron	
Steel	
Copper	
Plastic	
Wood	
No material (air)	

ⒽMagnetic force on a current-carrying wire: the motor effect

Figure 11.9 shows a length of copper wire that has been placed in a magnetic field. The copper is not magnetic so the wire itself is not affected by the magnet.

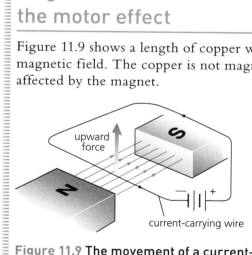

Figure 11.9 **The movement of a current-carrying wire in a magnetic field**

Fleming's left-hand rule

When a current flows there is a force on the wire. The force arises because the current produces its own magnetic field, which interacts with the field of the magnet. The resulting magnetic field is shown in Figure 11.10.

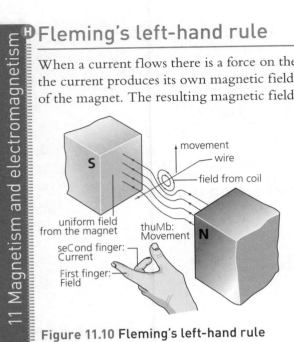

Figure 11.10 Fleming's left-hand rule

Note how the field lines between the poles of the magnet, which were originally straight, have become distorted due to the current in the wire. The result is that the wire experiences a force as the field lines tend to straighten.

If either the current in the wire changes direction or the polarity of the magnet is reversed, then the direction of the force on the wire is reversed. The force is increased if:
● the current in the wire is increased
● a stronger magnet is used
● the length of the wire exposed to the magnetic field is increased.

The relationship between the direction of motion, the current and the magnetic field when a current-carrying wire is in a magnetic field is predicted by **Fleming's left-hand rule**.

When applying this rule it is important to remember how the field and current directions are defined:
● The field direction is from the N pole of a magnet to the S pole.
● The current direction is from the positive (+) terminal of a battery to the negative (−) terminal.

Fleming's left-hand rule only applies if the current and field directions are at right angles to each other.

> **Fleming's left-hand rule** is the relationship between the direction of the current, the magnetic field and the force on a conductor

The turning effect on a current-carrying coil in a magnetic field

The loop in Figure 11.11 lies between the poles of a magnet. The current flows in opposite directions along the two sides of the loop. If you apply Fleming's left-hand rule, when a current is passed through the loop, one side of the loop is pushed up and the other side is pushed down. In other words there is a turning effect on the loop.

If the number of loops is increased to form a coil, the turning effect is greatly increased. This is the principle involved in electric motors.

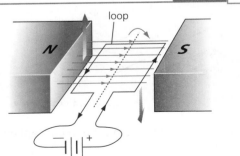

Figure 11.11 The force on a current-carrying loop in a magnetic field

⊕The d.c. electric motor

Figure 11.12 shows a simple electric motor. It runs on direct current from a battery. The coil is free to rotate on an axle between the poles of a magnet. The commutator, or split-ring, is fixed to the coil and rotates with it.

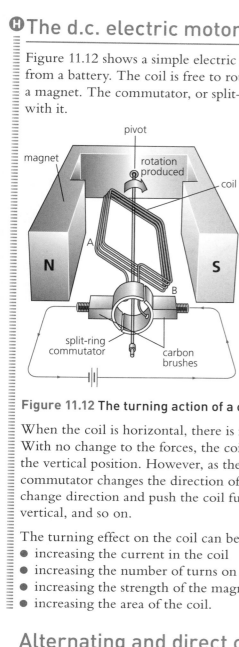

Figure 11.12 The turning action of a coil in a simple electric motor

When the coil is horizontal, there is maximum turning effect on the coil. With no change to the forces, the coil would eventually come to rest in the vertical position. However, as the coil overshoots the vertical, the commutator changes the direction of the current through it. So the forces change direction and push the coil further around until it is again vertical, and so on.

The turning effect on the coil can be increased by:
● increasing the current in the coil
● increasing the number of turns on the coil
● increasing the strength of the magnetic field
● increasing the area of the coil.

Exam tip

Details of the commutator are *not* required by the specification.

Alternating and direct currents

A direct current (d.c.) always flows in the same direction, from a fixed positive terminal to the fixed negative terminal of a supply.

A typical d.c. circuit is shown in Figure 11.13. A cell or battery gives a constant (steady) direct current. An oscilloscope trace of voltage versus time for a d.c. supply is shown in Figure 11.14.

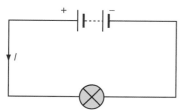

Figure 11.13 A simple d.c. circuit

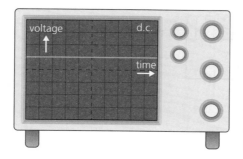

Figure 11.14 A trace of voltage against time for a d.c. supply

The electricity supply in your home is an alternating current (a.c.) supply. In an a.c. supply, the voltage (and hence the current) changes size and direction in a regular and repetitive way (Figure 11.15).

It is clear from Figure 11.16 why an a.c. supply is said to be bidirectional.

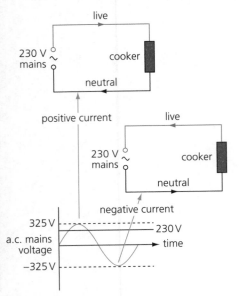

Figure 11.15 An a.c. supply

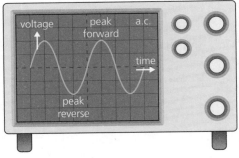

Figure 11.16 A trace of voltage against time for an a.c. supply

Now test yourself

3 A copper rod is positioned in an east–west direction and a plotting compass is placed at each end, as shown in Figure 11.17. North is vertically up the page.

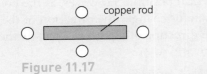

Figure 11.17

(a) Copy the diagram and mark on it the direction in which the arrow in the plotting compass will point in each of the four positions shown.

(b) The copper rod is now replaced by a bar magnet, as shown in Figure 11.18. Copy Figure 11.18 and on it mark the direction in which the arrow in the plotting compass will point in each of the three positions shown. The direction for the compass nearest the north pole of the magnet has already been done for you.

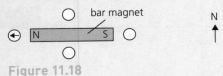

Figure 11.18

4 Two circuits are set up as in Figure 11.19. The iron rods are placed close together, and are free to move within the coils. Describe and explain fully what happens when the switches marked S are closed simultaneously.

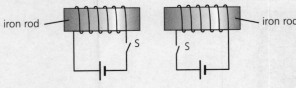

Figure 11.19

5 With the aid of a diagram, state Fleming's left-hand rule.

6 Use Fleming's left-hand rule to answer the following questions:

(a) Copy the diagrams in Figure 11.20. Draw arrows on parts (i) and (ii) to show the directions of the forces on the current-carrying wires.

⊗ Means current is perpendicular to and into the plane of the paper
⊙ Means current is perpendicular to but out of the plane of the paper

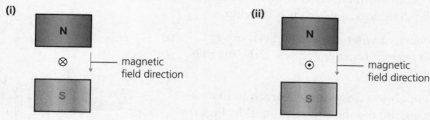

Figure 11.20

(b) Copy the diagrams in Figure 11.21 and indicate the direction of the current in the wire and the direction of the magnetic field in part (i). In part (ii), indicate the direction of the magnetic field and label the polarity of the magnets.

Figure 11.21

7 Figure 11.22 shows five displays on a cathode ray oscilloscope (CRO) screen. Which of the displays are:

(a) d.c. voltages

(b) a.c. voltages?

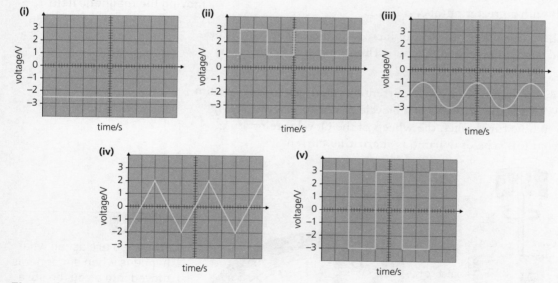

Figure 11.22

Electromagnetic induction

Electromagnetic induction (EMI) is the reverse of the motor effect. In EMI the user supplies a magnetic field and motion to produce a voltage that can cause a current.

The easiest way to demonstrate electromagnetic induction is with a magnet, a coil of wire and a centre–zero ammeter to measure the **induced** current and detect the direction in which it is flowing.

In Figure 11.23 a current is generated when the magnet is moving into or out of the solenoid. No current is generated when the magnet and coil are stationary.

Lenz's law states that the current always flows in the direction that will set up a magnetic field to oppose the motion causing it. In Figure 11.23(a), the current flowing in the coil produces a north pole at the left-hand end of the coil. As the magnet moves towards the solenoid, there is a magnetic force that repels it, so you have to do some work to push the magnet into the solenoid. The work done in pushing the magnet generates the electrical energy.

In Figure 11.23(b) the magnet is being pulled out of the solenoid. The direction of the current is reversed and so there is now an attractive force acting on the magnet. In Figure 11.23(a) we would get the same effect as moving the magnet to the right by moving the coil to the left. The important idea is that there must be *relative movement* between the magnet and the coil.

Faraday made three important discoveries about EMI. He found out that he could increase the size of the induced current by:
● moving the magnet faster
● using a stronger magnet
● using a coil with a greater number of turns.

All three of the above have the effect of increasing the rate at which the magnetic field (sometimes called magnetic flux) linked with the coil is changing.

It is also possible to induce an electric current by rotating a magnet within a coil of wire. This is the principle of a bicycle **dynamo** (Figure 11.24). The dynamo spindle turns when the wheels of the bicycle are rotating. The output from this type of dynamo is alternating current.

electromagnetic induction is the production of a voltage across an electrical conductor when the strength of the magnetic field around it changes

(a)
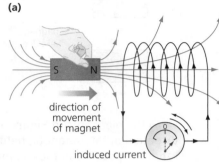
direction of movement of magnet
induced current

(b)
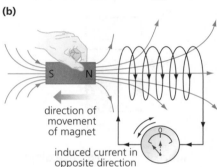
direction of movement of magnet
induced current in opposite direction

Figure 11.23 Inducing a current by moving the magnetic field

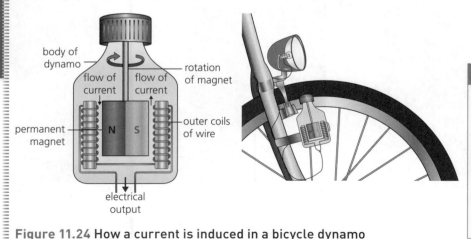
Figure 11.24 How a current is induced in a bicycle dynamo

Exam tip

When you are asked what happens when a magnet is moved into a coil, be sure to say that the current is momentary and then returns to zero. Remember, there is only a current while the magnet is actually moving.

F EMI can also be demonstrated by moving a single wire between the opposite poles of horseshoe magnet (Figure 11.25).

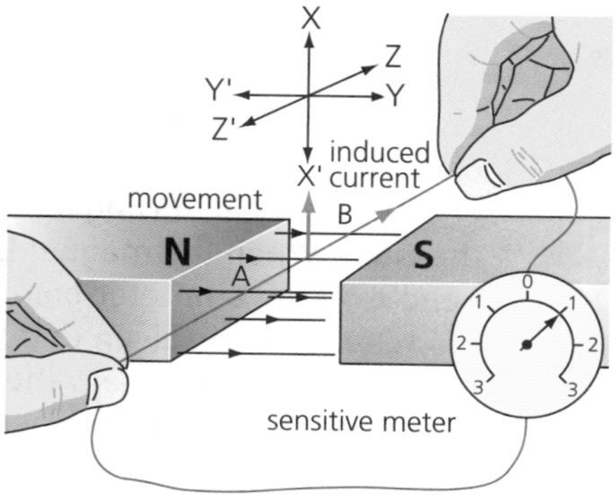

By carrying out the investigation shown in Figure 11.25 it can be shown that generation of the current depends on the direction of movement of the wire. To generate a current, the wire must cross the magnetic field lines. A current is produced if the wire is moved up and down along the direction XX′, but there is no current if the wire moves along ZZ′ or YY′. Reversing the direction of movement reverses the current. So, if moving the wire up makes the pointer in the meter move to the right, then moving the wire down will make it go to the left.

The size of the current generated can be increased in the following ways:
- moving the wire more quickly
- using a stronger magnet
- looping the wire so that several turns of wire pass through the poles.

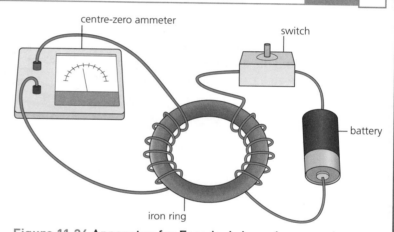

Figure 11.25 Electromagnetic induction using a single wire

Faraday's iron ring experiment

This is one of the classic experiments of the nineteenth century. Look at Figure 11.26. The soft iron ring in the centre has two coils wrapped around it. The coil on the right is connected to a switch and a small battery. The coil on the left is connected to a centre–zero ammeter.

The coils are made of **insulated** copper wire, so no current can flow in the iron core. In the situation shown in the diagram there is no current flowing in either coil.

What happens when the switch is closed? A current immediately flows in the coil on the right. But Faraday observed that there was a momentary deflection in the centre–zero ammeter. How could this be explained?

Figure 11.26 Apparatus for Faraday's iron ring experiment

Faraday argued:
- Closing the switch causes an increasing magnetic field in the right-hand coil.
- Since the iron core links the two coils magnetically, there is an increasing magnetic field in the left-hand coil causing an induced current, which flows through the ammeter.

The effect is momentary; there is a deflection of the ammeter needle only when the current in the right-hand coil is changing.

What then happens when the switch is opened? There is a momentary deflection of the ammeter needle, but now it is in the opposite direction because the field in the left-hand coil is now decreasing.

This simple experiment is the basis for the transformer, without which electrical energy could not be transmitted efficiently from power stations to our homes.

Now test yourself

8 (a) Look at Figure 11.27. If the magnet is stationary the reading on the centre-zero ammeter is zero. When the north pole of the magnet is moving into the coil, the needle moves to the right. When the magnet is stationary inside the coil, the needle on the ammeter points to zero.

   (i) What name is given to this effect?

   (ii) Explain these observations.

  (b) Describe what would be observed when the magnet is pulled up out of the coil.

  (c) What energy provides the source for the electrical energy in the coil?

9 Figure 11.28 shows the apparatus used in Faraday's iron ring experiment.

The coils wrapped around the iron ring are made of insulated copper wire.

  (a) Explain why a current flows in the secondary coil when the switch is closed.

  (b) Explain why a current flows in the secondary coil when the switch is then opened.

  (c) In what way is the current in the secondary coil different in parts (a) and (b)?

  (d) How could the apparatus be changed to produce an a.c. current in the secondary coil?

Answers online

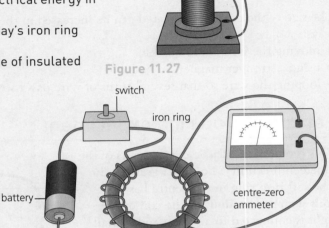

Figure 11.27

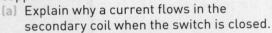

Figure 11.28

The a.c. generator

Figure 11.29(a) shows the design of a simple a.c. generator. It only produces alternating current, so it is called an alternator.

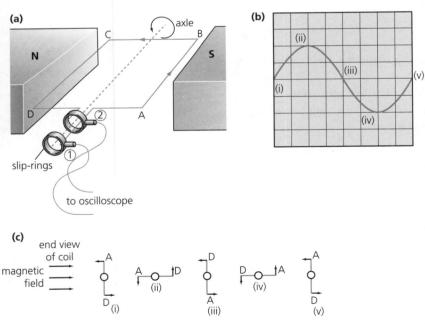

Figure 11.29 (a) An a.c. generator **(b)** How the voltage waveform produced by the generator appears on an oscilloscope screen **(c)** The position of the coil

> **Exam tip**
>
> Remember, in the motor effect, you supply a magnetic field and a current, and the result is a force that causes motion.
>
> In induction, you supply a magnetic field and motion (kinetic energy) and the result is an induced voltage or current.

F Turning the axle makes the coil rotate in the magnetic field. The coil is shown with a single turn to keep the diagram as simple as possible. The rotation of the coil causes a voltage to be induced across the ends of the coil.

As the coil turns the slip rings rotate. In contact with the slip rings are stationary brushes (marked 1 and 2 in the diagram). These provide a continuous contact between the coil and the cathode ray oscilloscope (CRO).

Figure 11.29(b) shows how the induced voltage changes with time. This can be observed on the CRO. Figure 11.29(c) shows the position of the coil at various times.

(i) The coil is vertical. Sides AB and CD are moving parallel to the field lines, so no voltage is induced.

(ii) The coil is horizontal. In this position sides AB and CD are cutting the field lines at the greatest rate. So the induced voltage is a maximum.

(iii) The coil is vertical once more, so there is no induced voltage.

(iv) The coil is horizontal once more. But notice that AB and CD are moving through the field in the opposite direction when compared with position (ii). So the induced voltage is once again a maximum, but now it is in the opposite direction.

(v) The coil is vertical once more, so there is no induced voltage.

The size of the induced voltage can be made larger by:

- rotating the coil faster
- using a coil of more turns
- using stronger magnets
- wrapping the coil around a soft iron core.

Transformers

REVISED ☐

A **transformer** works only with alternating current. Both the input and the output voltages are a.c.

Figure 11.30(a) shows the construction of a **step-up transformer**. All transformers consist of two coils of wire wrapped round a laminated iron core. A step-up transformer has fewer turns in the primary coil than in the secondary coil. It is used to increase the voltage.

Figure 11.30(b) shows the construction of a **step-down transformer** — one that has fewer turns in the secondary coil than in the primary coil. It is used to decrease the voltage.

Figure 11.30(c) shows the circuit symbol for a transformer.

> **step-up transformer** is a transformer in which the output voltage is greater than the input voltage (because the output coil has more turns than the input coil)
>
> **step-down transformer** is a transformer in which the output voltage is smaller than the input voltage (because the output coil has fewer turns than the input coil)
>
> **primary coil** is the input coil in a transformer
>
> **secondary coil** is the output coil in a transformer
>
> **laminated iron core** is the iron core of a transformer on which the input coil and the output coil are wound

(a)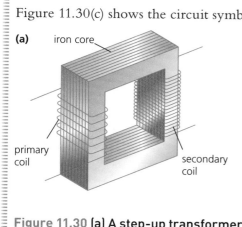

(b)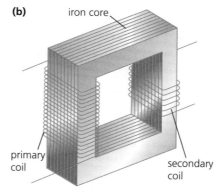

(c)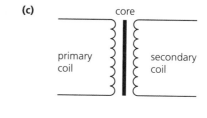

Figure 11.30 (a) A step-up transformer **(b)** A step-down transformer **(c)** The circuit symbol for a transformer

F All transformers have three parts:
- a **primary coil** — the incoming voltage, V_p (voltage across primary coil) is connected across this coil
- a **secondary coil** — this provides the output voltage, V_s
- a **laminated iron core** — this links the two coils magnetically.

H An important equation, known as the transformer equation, relates the two voltages V_p and V_s to the number of turns on each coil, N_p and N_s:

$$\frac{\text{number of turns in secondary coil}}{\text{number of turns in primary coil}} = \frac{\text{voltage across secondary coil}}{\text{voltage across primary coil}}$$

$$\frac{N_s}{N_p} = \frac{V_s}{V_p}$$

Power in transformers

REVISED

Transformers are very efficient devices with an efficiency that is typically above 99%. At GCSE we can treat transformers as being 100% efficient. For such a transformer we can write:

input electrical power to primary coil = output electrical power from secondary coil

$$V_p \times I_p = V_s \times I_s$$

where V_p = the voltage across the primary coil

I_p = the current in the primary coil

V_s = the voltage across the secondary coil

I_s = the current in the secondary coil.

F Electricity transmission

REVISED

Electrical power is **distributed** around the country from power stations through a grid of high-voltage power lines. The electricity in overhead power lines is **transmitted** to our homes and industry at 275 kV or 400 kV.

Although dangerous, high voltages are used because this means that there is then a low current in the cables and this wastes less energy. Before electricity reaches homes the high voltage must be stepped down to a much safer 230 V. Today almost all domestic electrical appliances in Europe operate at 230 V.

> **electricity distribution** is the passage of electric current from the end of the transmission system to the factories and homes that use it
>
> **electricity transmission** is the passage of electric current through the cables suspended from high pylons

Now test yourself

TESTED

10 A bar magnet is moving towards a loop of wire as shown in Figure 11.31.
   (a) Is a voltage induced in the wire? Give a reason for your answer.
   (b) Is a current induced in the wire? Give a reason for your answer.
   (c) The open ends of the wire are now connected to a torch bulb. Copy Figure 11.31, add the bulb and mark on your diagram the direction in which the current flows.

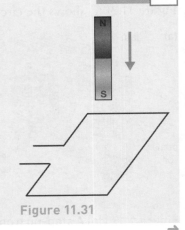

Figure 11.31

11 A coil of insulated copper wire, wrapped around a ring of soft iron is connected to a battery via a switch (Figure 11.32). A coil on the other side of the ring is connected to a centre-zero ammeter. Describe and explain what is seen on the ammeter when:

(a) the switch is closed

(b) the switch is re-opened.

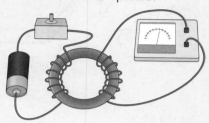

Figure 11.32

12 An anemometer is a device to measure wind speed. Figure 11.33 shows a simple anemometer. When the wind blows, the plastic cups turn.

(a) Explain why the wind causes the voltmeter to give a reading.

(b) Explain why the reading on the voltmeter is a 'measure' of the wind speed.

(c) The gauge is not sensitive enough to measure the speed of a gentle breeze. Suggest one way that the design can be modified to make the anemometer more sensitive.

13 At a power station, the main transformer is supplied from a 25 kV generator.

(a) How much energy is transferred from the generator for each coulomb of charge?

(b) The main transformer steps up the voltage to 275 kV before sending it out to the grid. Describe fully the purpose of stepping up the voltage.

(c) In what other part of the electricity transmission system must transformers be used?

(d) Why must these other transformers be used?

Answers online

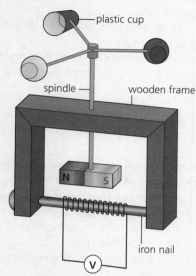

Figure 11.33

> **Exam tip**
>
> You can't predict what will come up in a physics exam, but it is a good bet that you'll get something on:
> - electromagnetism, or
> - the motor effect, or
> - electromagnetic induction, or
> - transformers.
>
> So learn the material well.

Exam practice

1 An electromagnet is a coil of wire through which a current can be passed.

(a) State three ways in which the strength of the electromagnet may be increased. [3]

(b) An electromagnet may be switched on and off. Suggest one situation where this would be an advantage over the constant field permanent magnet. [1]

(c) A coil carrying a current has two magnetic poles.

current

Figure 11.34

Copy Figure 11.34 and mark the magnetic poles produced. [2]
On your diagram, draw the magnetic field produced. [4]

→

2 A taut straight wire XY is placed between the opposite
poles of a magnet. When an electric current flows in the
wire XY, it experiences a force which is vertically upwards
(Figure 11.35).
  (a) Use an arrow to show the direction of the
    current in XY. [1]
  (b) Complete the circuit by adding the symbol
    for a battery, having due regard to its
    polarity. [1]

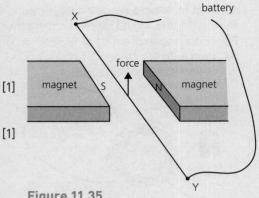

Figure 11.35

3 A metal rod AB is placed near a magnet. End A is strongly
attracted when it is placed near the north pole or near the
south pole of the magnet (see Figure 11.36).
  (a) Describe and explain what you would observe when
    end B is placed, in turn, near the north and south
    poles of the magnet. [4]
  (b) From what metal is the rod most likely to
    be made? [1]

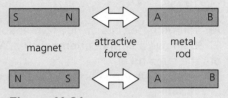

Figure 11.36

4 (a) What is a plotting compass? [1]
  (b) Figure 11.37 shows a bar magnet. Describe how could
    you use a plotting compass to show the magnetic field
    pattern around the magnet. [3]
  (c) Copy Figure 11.37 and draw the magnetic field
    around it. [3]

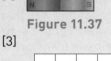

Figure 11.37

5 (a) What is the difference between alternating current
    (a.c.) and direct current (d.c.)? [2]
  (b) On a sheet of graph paper similar to that in
    Figure 11.38, sketch the waveform of:
    (i) an a.c. signal [1]
    (ii) a steady d.c. signal [1]
    (iii) a changing d.c. signal in which the current
      direction is opposite to that in part (ii) [1]
  (c) State a source for (i) a.c. and (ii) d.c. [2]

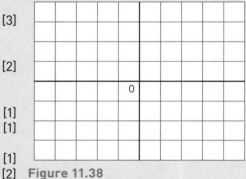

Figure 11.38

6 Figure 11.39 shows a coil of wire wrapped around a cardboard
tube. When the switch is closed, a current flows, creating a
magnetic field.
  (a) Copy Figure 11.39 and on it draw five field lines representing
    the shape of the field pattern. Show the direction of all field
    lines with arrows. [3]
  (b) Use the letter N to mark on the diagram the end of
    the tube that behaves as a north pole. [1]
  (c) State three ways to increase the strength of the field
    in the coil. [3]

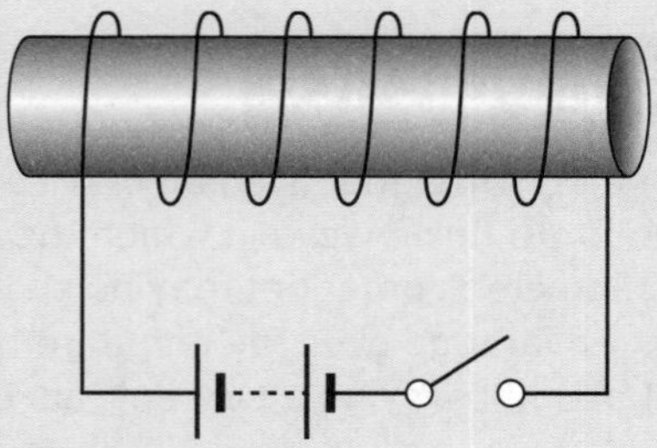

Figure 11.39

7 Figure 11.40 shows two coils of wire placed close together. One coil is connected to a centre-zero ammeter and a resistor. The other coil is connected to a battery and a switch.

(a) Describe carefully what you would observe on the centre-zero ammeter when the switch is closed. [2]

(b) What is observed on the centre-zero ammeter when the switch is re-opened? [2]

(c) What name is given to this effect? [1]

(d) How could the size of the effect on the ammeter be increased without changing the battery, the coils or the resistor? [1]

(e) Suggest a reason for including a resistor in series with the ammeter. [1]

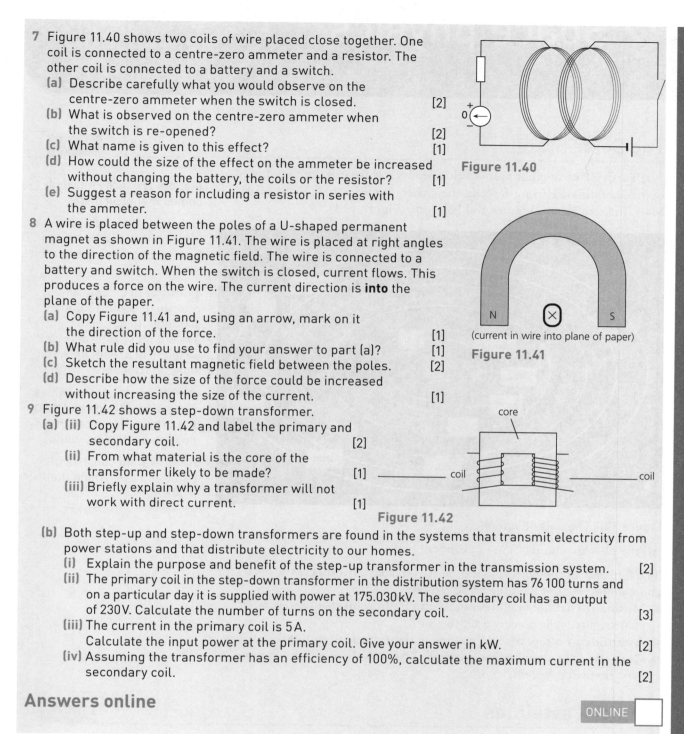

Figure 11.40

8 A wire is placed between the poles of a U-shaped permanent magnet as shown in Figure 11.41. The wire is placed at right angles to the direction of the magnetic field. The wire is connected to a battery and switch. When the switch is closed, current flows. This produces a force on the wire. The current direction is **into** the plane of the paper.

(a) Copy Figure 11.41 and, using an arrow, mark on it the direction of the force. [1]

(b) What rule did you use to find your answer to part (a)? [1]

(c) Sketch the resultant magnetic field between the poles. [2]

(d) Describe how the size of the force could be increased without increasing the size of the current. [1]

(current in wire into plane of paper)

Figure 11.41

9 Figure 11.42 shows a step-down transformer.

(a) (ii) Copy Figure 11.42 and label the primary and secondary coil. [2]

(ii) From what material is the core of the transformer likely to be made? [1]

(iii) Briefly explain why a transformer will not work with direct current. [1]

Figure 11.42

(b) Both step-up and step-down transformers are found in the systems that transmit electricity from power stations and that distribute electricity to our homes.

(i) Explain the purpose and benefit of the step-up transformer in the transmission system. [2]

(ii) The primary coil in the step-down transformer in the distribution system has 76 100 turns and on a particular day it is supplied with power at 175.030 kV. The secondary coil has an output of 230 V. Calculate the number of turns on the secondary coil. [3]

(iii) The current in the primary coil is 5 A.
Calculate the input power at the primary coil. Give your answer in kW. [2]

(iv) Assuming the transformer has an efficiency of 100%, calculate the maximum current in the secondary coil. [2]

Answers online

ONLINE ☐

12 Space physics

The Solar System

The Earth is one of eight **planets** that orbit a star we call the Sun. Listed in order of distance from the Sun, the planets are: Mercury, Venus, Earth, Mars, Jupiter, Saturn, Uranus and Neptune (Figure 12.1).

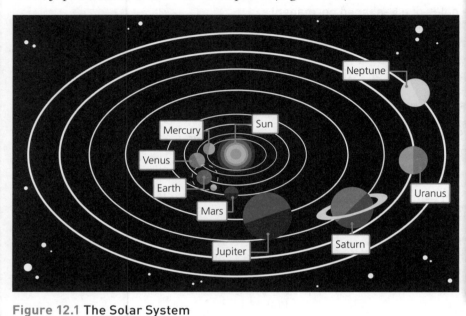

> **planets** are large heavenly bodies that orbit a star
>
> **elliptical** is the oval shape of the path taken by many heavenly bodies as they orbit the Sun
>
> **asteroids** are lumps of rock up to about 1000 km diameter. In our Solar System most asteroids can be found in a 'belt' between Mars and Jupiter
>
> **comets** are heavenly bodies made of mostly ice and dust that orbit a star, often in very elongated elliptical paths
>
> **Solar System** consists of a star and everything that orbits it

Figure 12.1 The Solar System

All the planets orbit the Sun in **elliptical** paths.

The Sun and all the objects that orbit it — planets, **moons**, **asteroids** and **comets** — are called the **Solar System**.

- **Moons** are natural satellites of planets. This means that moons orbit planets, just as the planets orbit the Sun.
- **Asteroids** are lumps of rock, ranging in size from 1 km to 1000 km across.
- **Comets** are made up of ice and dust. They travel around the Sun in very elongated (eccentric) orbits.

Artificial satellites

The Moon is a **natural satellite** of the Earth. Since the late 1950s, people have put **artificial satellites** into orbit around the Earth. They are used mainly for:
- astronomy (for example the Hubble telescope)
- communications (long-distance phone calls and radio/television broadcasts)
- weather monitoring/forecasting
- monitoring agricultural land use
- monitoring military activity, and general espionage.

Now test yourself

1 Which rocky planet is furthest from the Sun?
2 Which gas planet is nearest to the Sun?
3 In 2006 the USA launched a probe to explore the outer reaches of our Solar System. Why was collision with an asteroid much more likely than collision with a comet?
4 The Hubble telescope is an *artificial satellite*. What does this mean?
5 What name is given to the natural satellites that orbit the planets?

Answers online

The life cycle of stars

Star birth

A star is formed from clouds of hydrogen and dust, known as **stellar nebulae**. The force of **gravity** causes particles of **hydrogen** to come together. These clouds become more and more **dense** as the particles get closer and closer together. The hydrogen particles start to spiral inwards — a process called **gravitational collapse**.

The temperature rises enormously and the hot core at the centre is called a **protostar**. When the temperature at the core reaches about 15 million °C, **nuclear fusion** begins and a star is born.

The outward radiation pressure is balanced by the inward gravitational force, so the size of the star remains stable. The star is now in the main phase of its life, so it is called a **main sequence** star.

Ⓗ Death of a star like our Sun

When almost all of the hydrogen is used up in fusion, the energy output reduces so much that gravity compresses the star significantly, but the star doesn't shrink. Surrounding the core is a layer of hydrogen. Gravitational contraction provides enough energy for nuclear fusion of the hydrogen in *this* layer. The outward pressure from the nuclear fusion reactions prevents the star from collapsing and makes it *expand* to several hundred times its former size. The surface temperature falls and the starlight is now predominantly orange. The star is now called a **red giant**.

Other nuclear reactions now take place within the red giant. Helium, for example, fuses to become carbon and oxygen. Close to the end of the life of a red giant the gravitational force can no longer hold the outer layers of gas. These gases flow out and cool to form a nebula. This nebula may eventually contribute to the creation of another star. Over time the core that remains cools to become a **white dwarf**.

Eventually, all fusion stops, and the star cools further to become a **black dwarf**.

---

stellar nebula is a cloud of gas and dust from which stars are formed

gravitational collapse is a process in the evolution of a star in which hydrogen particles get closer and closer together because of gravity

protostar is a very, very hot ball of gas in which nuclear fusion has not yet begun — when fusion begins the protostar becomes a star

main sequence is the most stable stage in the mid-life of a star

red giant is a star that has used up almost all of its hydrogen and the outward pressure from nuclear fusion makes it expand to several hundred times its normal size

white dwarf is one of the last stages in the life cycle of a star like our Sun

black dwarf is the final state of a star, like our Sun, when all fusion stops and it becomes very cold

Death of a high-mass star

In a star of high mass the rate at which helium fusion occurs is much more rapid than for a star of smaller mass. The huge amount of energy from helium fusion pushes the outer layers of the star outwards and it turns into a **red supergiant**.

Red supergiants burn through their nuclear fuel very quickly and most live for only a few tens of millions of years. A red supergiant successively fuses in its core different elements in the periodic table, up to the creation of iron. At that point the supergiant begins to collapse. This collapse releases gravitational potential energy that heats up and throws off the outer layers of the star in the form of an enormous explosion called a **supernova**. For about a month the supernova emits more radiation than all the other stars in its galaxy put together!

The core of the star is all that is left, an unimaginably dense object called a **neutron star**. Neutron stars are the smallest and densest stars known to exist, typically with a radius of about 10 km. In the most massive of stars a **black hole** is created. Black holes have such enormous gravitational fields that *nothing* can escape from them — not even light. That is why we call them *black* holes.

The life cycle of a star is illustrated in Figure 12.2.

> **red supergiant** is a stage in the life cycle of a very massive star following the main sequence stage
>
> **supernova** is a stage near the end of the life cycle of a very massive star when it suddenly increases greatly in brightness because of a catastrophic explosion that ejects most of its mass
>
> **neutron star** is the collapsed core of a large star
>
> **black hole** is the collapsed core of a very large star. It is so called because the gravitational force is so great that not even light can escape from it

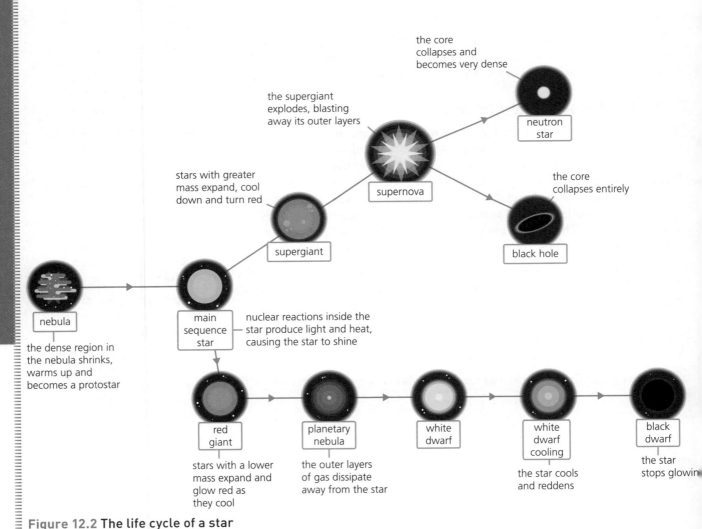

Figure 12.2 The life cycle of a star

Now test yourself

6 What is a stellar nebula?
7 Where does the energy come from to change a stellar nebula into a protostar?
8 Not all protostars become stars, even though they are hot enough to emit light. Suggest a reason why not.
9 What process produces heat energy in stars like our Sun?
10 What is the approximate temperature at the core of our Sun?
11 Our Sun has remained the same size in the sky for about 4 billion years because the forces on it are balanced. What balances the inward gravitational force?
12 Why will our Sun become a red giant when it ceases to be a main sequence star?
13 What will be the final fate of our Sun?
14 What stage in the life cycle of a very massive star comes immediately before a supernova?

Answers online

Planet formation

REVISED

During a period of gravitational collapse, the hydrogen nebula that eventually forms a protostar starts to spin. It is generally thought that this spinning disc of gas and dust is 'blown away' by the star once fusion begins. This nebula is now called a **planetary nebula**.

Over many millions of years the dust and gas come together as a result of gravity. Eventually small rocks coalesce (or **accrete**) to become larger rocks and so on until they emerge as planets. This theory explains why all the planets in our Solar System orbit the Sun in the same sense and are found in the same plane — they were all formed from the same spinning nebula.

> **planetary nebula** is a cloud of gas and dust from which planets are made

The galaxies

REVISED

Our Solar System contains only one star — the Sun, but as we look into the night sky we see a vast number of star systems. These make up our **galaxy**, the **Milky Way**.

A typical galaxy contains around a billion stars and the Universe is thought to contain over a hundred billion galaxies.

> **galaxy** is a huge number of star systems held together by gravity
>
> **Milky Way** is a galaxy consisting of around 100 — 400 billion star systems, one of which is our own Solar System

Formation and evolution of the Universe

REVISED

Most physicists today accept the **Big Bang theory**. The Big Bang occurred around 14 billion years ago from a tiny point that physicists call a singularity.

Not long after the Big Bang, the Universe was made up of high-energy radiation and elementary particles like quarks, the particles that make up protons and neutrons. This was a period of rapid expansion or 'inflation'. Rapid expansion is always associated with cooling, so as the Universe got bigger it cooled down. This allowed the quarks to come together to form protons and neutrons.

> **Big Bang theory** is the theory that the Universe started around 14 billion years ago from a point known as a singularity

Further expansion and cooling allowed the temperature to fall sufficiently to enable electrons to combine with neutrons and protons to form atoms of hydrogen.

Red-shift

If a source of waves is moving, the crests of its waves get bunched together in front of the wave source. If the wave crests are bunched together, their wavelength decreases.

On the other side of the source (behind it), the waves spread out and the wavelength increases.

This is why the sound of a siren from an emergency vehicle appears to have a higher pitch (smaller wavelength) as it approaches us and a lower pitch (bigger wavelength) as it moves away from us. This is called the **Doppler effect**.

Figure 12.3(a) shows a source of sound at rest. In Figure 12.3(b) the sound is moving to the right. Observer A will hear high-pitch sound (shorter wavelength) as the sound waves are being bunched together. Observer B hears low-pitch sound (longer wavelength) as the waves are being spread out.

A similar thing occurs with light. Visible light consists of much more than seven different colours. Physicists prefer to think that there is a continuum from red to violet — so there are an infinite number of colours in the visible spectrum. Each colour has a wavelength associated with it.

If the light that we observe from a moving source has a shorter wavelength than expected, it is because the source is moving towards us — we say the light is 'blue-shifted'. But if the light we observe has a longer wavelength than expected, it is because the source is moving away from us — and we say the light is '**red-shifted**'.

Our Sun contains hydrogen. We know this because there are black lines (known as Fraunhofer lines) in the spectrum of the light from the Sun, where hydrogen atoms have absorbed light. This pattern of black lines is called the absorption spectrum for hydrogen (Figure 12.4).

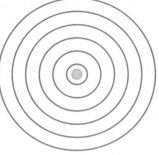

(a) source of sound at rest

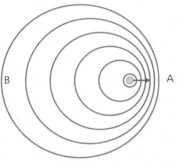

(b) source of sound moving to the right

Figure 12.3 The Doppler effect

> **red-shift** is the increase in the wavelength of light from distant galaxies due to their increasing separation from us

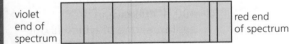

violet
end of
spectrum

red end
of spectrum

Figure 12.4 Absorption spectrum for hydrogen

By closely examining the light spectrum, physicists have identified over 50 different elements in the Sun.

What happens when we look at the light from distant galaxies?
● We get the same pattern as we do from the Sun but it is shifted towards the red end of the spectrum.
● The fact that we always get red-shift from the distant galaxies tells us that the galaxies are all moving away from us.
● This tells us that the Universe is *expanding* — the distance between the galaxies is getting bigger and bigger.

An expanding Universe supports the Big Bang theory.

The spectra in Figure 12.5 demonstrate red-shift and show that both Nubecula and Leo are moving away from us.

Answers at **www.hoddereducation.co.uk/myrevisionnotesdownloads**

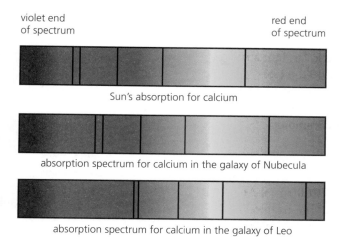

violet end
of spectrum

red end
of spectrum

Sun's absorption for calcium

absorption spectrum for calcium in the galaxy of Nubecula

absorption spectrum for calcium in the galaxy of Leo

Figure 12.5 Absorption spectra from different galaxies

Cosmic microwave background radiation

REVISED

In the 1960s two American physicists, Arno Penzias and Bob Wilson, discovered microwaves coming from all parts of the sky. Today most physicists believe that this continuous, **cosmic microwave background radiation (CMBR)** is the remnant or 'echo' of the Big Bang. The CMBR corresponds to the radiation emitted by a black body at a temperature of about −270°C or 3 kelvin. It is sometimes called 3K continuous background radiation.

The Big Bang theory is currently the only model that explains CMBR.

> **cosmic microwave background radiation (CMBR)** is now thought of as a 'signature' or 'after-glow' of the Big Bang

Exoplanets

REVISED

By the end of 2016, over 3600 planets had been discovered in over 2600 Solar Systems. They are called **exoplanets** because they are outside our Solar System ('exo' means outside). Of course, it is not known if any of these planets can support life as we know it.

There are two methods of detecting planets outside our Solar System. Both involve observing the light coming from stars that are similar to our Sun.

One way is to look for a 'transition' — a tiny reduction in the light reaching us from that star when an orbiting planet passes between the star and us.

Another way is to observe that the star and the planet go around a common centre of mass — this means that sometimes the star is moving towards us, and sometimes it is moving away from us. This in turn leads to tiny variations in the wavelengths of the light from that star due to the Doppler effect.

> **exoplanet** is a planet outside of our Solar System

Planetary atmospheres

REVISED

Of particular importance to physicists studying planetary atmospheres is the existence of oxygen and water. Oxygen and water are essential for human life and their presence in the atmosphere of any planet might indicate the possibility of life there.

Space travel within our Solar System

People first set foot on the surface of the Moon in 1969 and, so far, humans have never ventured further than the Moon. On 15 April 2010 the US President said: 'By the mid-2030s, I believe we can send humans to orbit Mars and return them safely to Earth. And a landing on Mars will follow.' This still remains an objective for many scientists in the astrophysics community.

Space travel beyond the Solar System

Our fastest spacecraft can travel at a maximum speed of 70 000 m/s. At this speed it would take a staggering 18 000 years to reach the nearest planet outside the Solar System. The vast distances to the stars mean that it is certain that with our present technology it is not feasible to visit any planet outside our Solar System. There are enormous difficulties:

- flight time — the distance is so great that the flight would last for many generations
- engineering — our spacecraft are just too slow
- logistics — it is not clear how the spacecraft could carry enough fuel, oxygen and water
- ethical — the chance of failure would be high, with no possibility of return to Earth.

H The speed of light

Most people are aware that light is incredibly fast. But just how fast is it? There is nothing known to travel faster than light. Light travels at 300 000 km/s (3×10^5 km/s or 3×10^8 m/s).

Light is very fast, but astronomical distances are also enormous. That is why astronomers sometimes measure distances in light years.

A light year (ly) is the distance light travels in 1 year.

So, just how far is 1 ly?

distance = speed × time = 3×10^8 m/s × $(60 \times 60 \times 24 \times 365)$ s

$\qquad = 9.46 \times 10^{15}$ m

$\qquad = 9.46 \times 10^{12}$ km

$\qquad$ = over 63 000 times the distance from the Earth to the Sun

> **Exam tip**
>
> If you start writing 'A light year is the time taken for...' your examiner will probably give you no marks, no matter what comes next. This is because a light year is a *distance*, not a time.

Now test yourself

H 15 What evidence is there to believe that there are elements like helium and calcium in our Sun?

16 What do physicists mean by red-shift in the context of astronomical observations?

17 What does red-shift tell us about our neighbouring galaxies?

18 What do the letters CMBR stand for?

19 CMBR was discovered by Penzias and Wilson. What explanation is put forward to explain it?

F 20 What are exoplanets?

21 Suggest why astrophysicists are so keen to discover whether there is oxygen and water on exoplanets.

Answers online

> **Exam tip**
>
> Your examiner *has* to ask some questions that require continuous prose (QWC questions).
>
> This chapter has lots of material suitable for this type of question:
> - what makes up our Solar System
> - life cycle of stars
> - supernovae, neutron stars and black holes
> - Big Bang model
> - red shift and CMBR

Exam practice

1 The photograph shows a cloud of gas and dust known as a nebula. The bright spots are stars.

(a) What force causes the gas to form stars? [1]
(b) What two gases are the main constituents of stars? [1]
(c) How do astronomers know this? [1]
(d) Name the process that supplies the energy in stars. [1]
(e) Apart from producing energy in stars what else is produced by this process? [1]

2 (a) Nuclear fusion of hydrogen is the main source of energy in our Sun. Suggest a reason why this requires enormously high temperatures to occur. [2]
(b) Suggest why stars which are much more massive than our Sun must use their fuel very rapidly if they are to remain stable. (Hint: what is the condition for stellar stability?) [3]
(c) Our Sun has a mass of 2×10^{30} kg. Its mass reduces by 4×10^{9} kg every second as a result of nuclear fusion. At the moment, by what percentage does the mass of the Sun decrease every year? [4]

3 Arrange these heavenly bodies in order of size (mass), beginning with the smallest:
asteroid   black hole   galaxy   neutron star   our Sun   planet [3]

4 (a) What name is given to our galaxy? [1]
(b) What is a light year? [2]
(c) The Small Magellanic Cloud is a galaxy about 200 000 light years away from us. If the speed of light is 3×10^{5} km/s, how far away from us is this galaxy in km? [3]

Answers online

ONLINE